VOLUME II: ROADSIDE AND RIVERSIDE GEOLOGY OF ILLINOIS

Steven D.J. Baumann, P.G., Teresa Arrospide, and Jamie L. Bardwell

Edition 2: ©2021

Midwest Institute of Geosciences and Engineering

(inside cover)

We would like to personally thank the following individuals who have made this book possible by either joining us in the field, selecting the stops, editing, or just by encouraging my love for rocks.

I dedicate this book to all of you!

David H. Malone

Sarah M.H. Baumann

Micaela Krol

Alexandra B. Cory

David Johnson

Mary M. Pryjda

Tiffany E. Mikes

Cover photo taken by Steven Baumann. Maps adapted from the Illinois State Geologic Survey's online maps. "Bedrock Geologic Map of Illinois" (2005) and "Quaternary (Ice Age) Deposits" (2005).

VOLUME II
ROADSIDE AND RIVERSIDE GEOLOGY OF ILLINOIS

STEVEN D.J. BAUMANN, P.G.
TERESA ARROSPIDE
JAMIE L. BARDWELL

Midwest Institute of Geosciences and Engineering

© 2017 and 2021

www.mige-web.org

EDITED BY: **Sandra K. Dylka**

FIGURES AND DIAGRAMS: **Teresa Arrospide**

PHOTOS TAKEN BY: **Steven D.J. Baumann**

From the primary author, Steven D.J. Baumann:

Thank-you for purchasing this book!

NOTES:

Table of Contents

Notes..........iii

Table of Contents..........iv

Preface for Volumes I and II..........1

Introduction 1

Stop Location Map..........2

Volume II: Roadside and Riverside Geology of Illinois 4

Northwestern Illinois..........5

    Driftless Section..........5

    Rock River Hill Country..........5

    Green River Lowland..........5

Guide for Northwestern Illinois..........5

    IL-2: Rockford to Dixon..........5

    US-20: Rockford to East Dubuque..........9

    I-39: Rockford to I-88..........18

    IL-84: The IL-84 and US-20 Junction to Savanna..........19

Northeastern Illinois..........22

    Chicago Lake Plain..........22

    Wheaton Morainal Country..........22

    Kankakee and Bloomington Ridged Plains..........23

Guide for Northeastern Illinois..........23

    IL-137 and US-12: Lake Michigan to McHenry..........23

    IL-31, IL-25, and IL-62: McHenry to Aurora..........25

    US-34: Aurora Area..........28

    I-90/I-94, I-80/I-294: Chicago Area..........30

    I-80, I-55, and I-355 Corridor: Naperville to Lemont to Channahon..........33

    IL-83: Lemont to Alsip..........38

    US-52 Corridor: Joliet to I-39..........39

    IL-1: Near the Indiana Border..........40

    IL-102 and IL-103: Wilmington to Kankakee..........41

    US-6, IL-71, IL-178, and IL-351: Ottawa to Oglesby..........42

Table of Contents

Western Illinois……….49

    Galesburg Plain……….49

    Griggsville Plain……….49

    Dissected Tills Plain……….49

    Lincoln Hills……….49

Guide for Western Illinois ……….49

    US-6 and I-88: Moline Area……….49

    I-74/747: Peoria Loop……….51

    IL-95 and US-24: Peoria to Mt. Sterling……….52

    IL-104 and I-72: Liberty to Griggsville……….57

Central Illinois ……….62

    Ancient Illinois Floodplain……….62

    Springfield Plain……….62

    Kankakee Plain………..62

    Bloomington Ridged Plains……….62

Guide for Central Illinois……….63

    IL-18 and IL-26: Henry Area East of the Illinois River……….63

    I-55: Bloomington to Springfield……….65

    I-57: Kankakee to Champaign……….66

    I-74: Bloomington to the Indiana Border……….69

    IL-100: Northern East St. Louis Area………..70

Table of Contents

Southern Illinois………..71

    Mount Vernon Hill Country……….71

    Shawnee Hills……….71

    Coastal Plain Province……….71

    Salem Plateau……….71

Guide for Southern Illinois……….72

    I-57: Mt. Vernon to Cairo……….72

    US-51: Carbondale to Anna……….74

    I-24: Goreville to Metropolis……….76

    IL-1 and IL-146: The Ohio River to Vienna……….81

    IL-34: Rosiclare to Harrisburg……….84

    IL-3: Cairo to East St. Louis……….86

Glossary……….97

Additional Reading……….104

References……….105

Disclaimer……….111

Preface for Volumes I and II

The stops in this book are either on major roadways or slightly off of them in streams and state parks. The stops included in this book are by no means all of the outcrops or locations of geologic interest in the state. Many of the locations in this book are along the interstate highways. In Illinois, it is legal to park along the interstate and explore the rocks. The only exception is along toll roads. They have very little geology exposed as it is. In Illinois, most exceptionally well exposed rocks are on private land and in fenced off quarries. There are some locations herein on private land. If a location is on private property, we attempted to only include places that are visible from a public vantage point. Please remember to ask permission before entering private property.

In order to start from any location, global positioning system, (GPS) coordinates are given in decimal format for precise locations. Photos are included to better orientate you along the roads or off the road. Northing is listed before the easting throughout the book. If you enter the GPS coordinates into your smartphone or vehicle GPS, it will take you to the spot.

The best times of the year to look at the geology of Illinois is late winter to late spring after the snow has melted and before the trees grow their leaves. Fall and early winter are also excellent times to view the geology. The 2017 edition was combined with the now separated out first volume called Geologic History of Illinois. This was done to lower the cost as some information has been added. This volume alone contains almost one hundred forty-eight color photos and five color diagrams.

I hope you find this book informative and enlightening. Illinois is a remarkable state and is so much more than a land of mega farms and cities.

Introduction

How to use this volume

Giving outcrop locations via directions and road maps has its limitations. Roads and railroads change. Buildings get repurposed or demolished. Things become more overgrown during the summer. Saying things like "turn right down the dirt road where you see the yellow barn", doesn't mean much decades years after you print that. Buildings get repainted or decay. Roads become paved, etc. But pre-GPS that was the only real way to do it. Fortunately our phones and GPS navigation devices will allow you to type in the GPS coordinates given in this volume. Type it in as it appears in this book, in decimal format. Don't forget to type the minus for westing. For example if you wanted to find the site on page 46 you would enter the coordinates "**41.3415, -89.0237**", just as it appears in this book. See below for a depiction of what to enter

The statewide maps herein are meant to be used as general guides and are in no way meant to represent the exact location to any accuracy. That's what the GPS is for. They are included to give you a general sense of an outcrop location, incase it is near you.

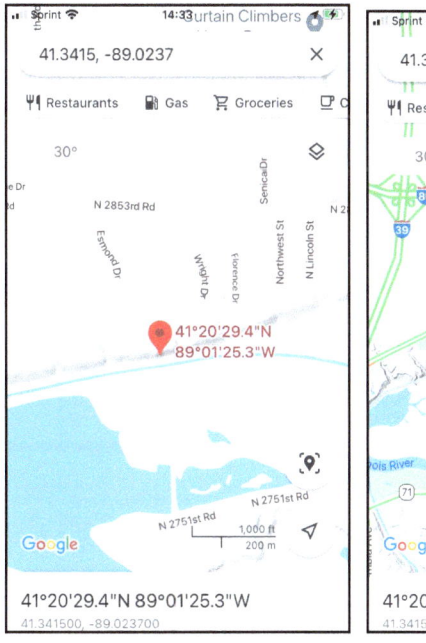 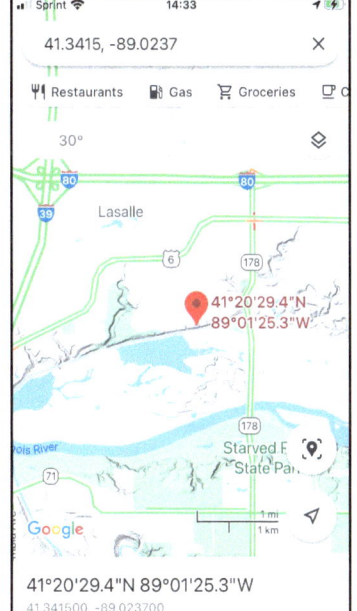 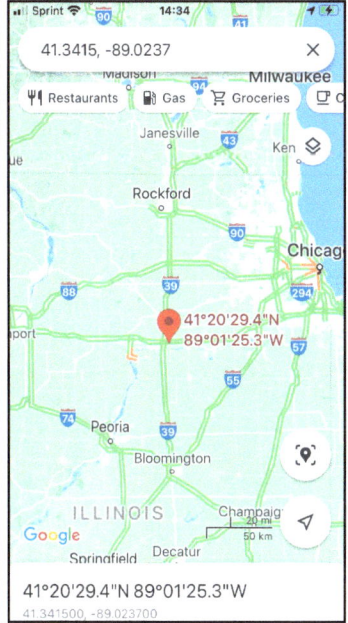

Stop Location Map

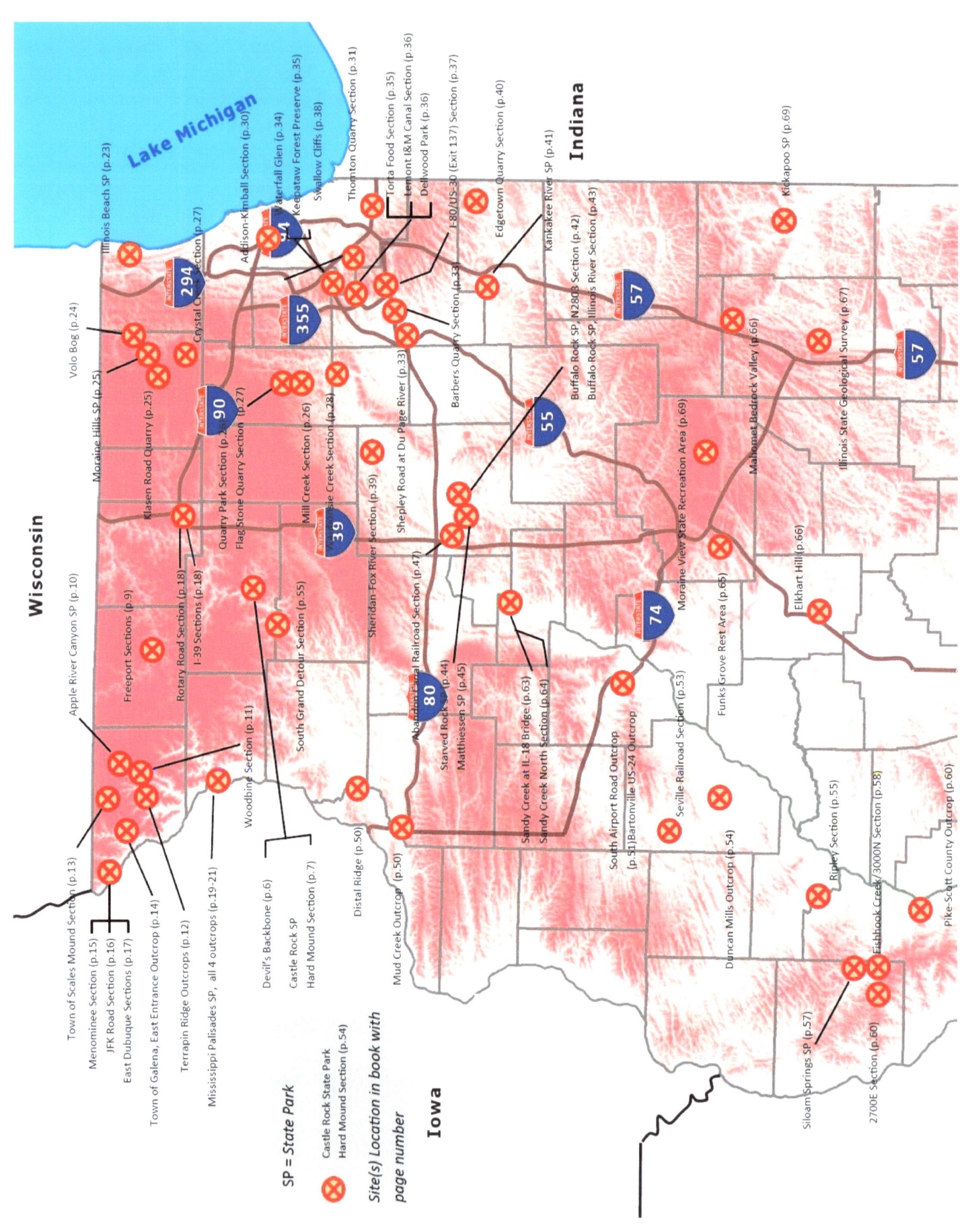

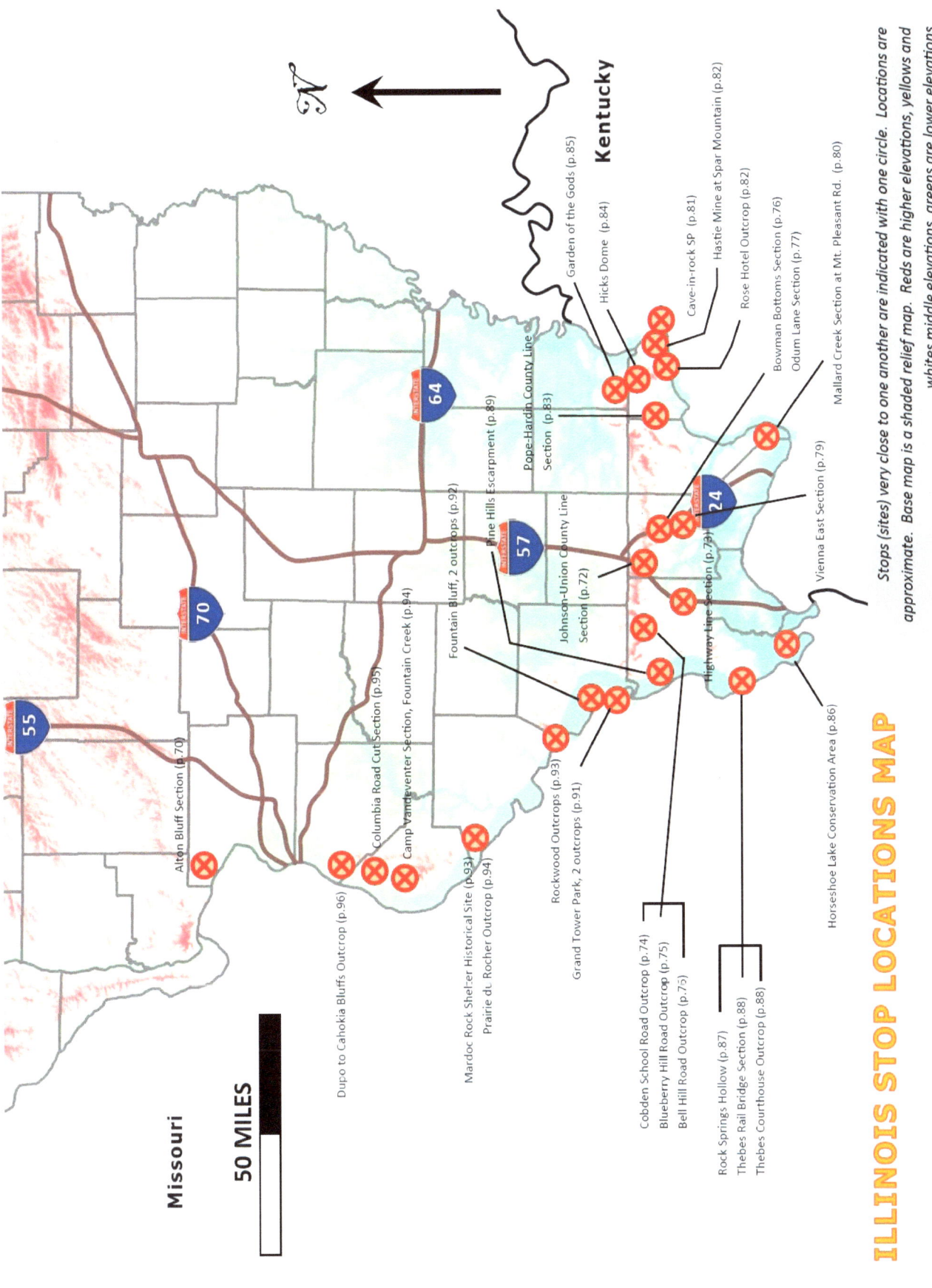

VOLUME II
Roadside and Riverside Geology of Illinois

Remember to enter the GPS coordinates (as they appear in this book) into your smartphone or navigation device.

Northwestern Illinois

Driftless Section

The Driftless Section is an area that covers almost all of JoDaviess County and the northwestern part of Carroll County. It extends north into Wisconsin's Baraboo area and has the rough shape of an inverted triangle that extends north to the Baraboo Wisconsin area. As the name implies, the Driftless Section has no glacial drift present within it. There are some loess and slack water lake deposits but the Quaternary ice sheets never covered the area. The areas surrounding the Driftless Section were all covered with ice. Geologists aren't quite sure why the glaciers avoided this area. Elevation, river patterns, and weather may have been some of the controlling factors. The Driftless Section is characterized by high hills and cliffs with deep and well-developed river valleys. Most of the good bedrock exposures in the area are the Middle Ordovician carbonates of the Galena and Platteville Groups. Late Ordovician shales and Silurian carbonates are also common. The major geologic structure in the area is the Plum River Fault Zone which enters the area from Iowa.

Rock River Hill Country

The Rock River Hill Country is the area between the Driftless Section and the Wheaton Morainal Country. It is characterized by its well-developed streams, rolling hills, thin glacial cover, and scattered bedrock outcrops. The Rock River Hill Country is mostly covered by glacial deposits of the Illinois Episode and Wisconsin Episode outwash. The bedrock in the area ranges in age mostly from Cambrian to Ordovician. The Wisconsin Arch gives way to the Kankakee Arch in the area. The Sandwich Fault Zone runs the length of the area. The northern portion of the LaSalle Anticlinorium begins here.

Green River Lowland

The Green River Lowland is one of the smaller physiographic divisions in Illinois covering about 800 square miles. This small, yet significant area has been shaped largely by wind. The major glacial deposits in the area consist of Wisconsin Episode outwash deposits of the Henry Formation overlying a thin mantle of Illinois Episode glacial drift. As the glaciers left the area high winds blew across the barren surface and deposited vast sand dunes. The largest are just south of the village of Hooppole and north of the Green River, along Illinois Route 78, about 5 miles north of I-80. The sand dunes are mostly covered with vegetation; they form a stark contrast standing as high hills relative to the surrounding landscape.

The bedrock in the area is mostly covered by glacial deposits and is similar to the bedrock of the Driftless Section and Rock River Hill Country. The western edge of the Sandwich Fault Zone resides in the area as does the LaSalle Anticlinorium.

Guide for Northwestern Illinois

IL-2: Rockford to Dixon

To explore the geology of Northwestern Illinois, there is perhaps no better launching point than Rockford. From Rockford you can head in a southern to western direction and see some of the best geology exposed in the state. The journey south from Rockford to Dixon along IL-2 is about a 40 mile trip that parallels the Rock River almost the entire way within the Rock River Hill Country. It is a landscape shaped heavily by the Illinois and Wisconsin glaciers. The trip also exposes bedrock close to the Rock River that has been altered by tectonic activity. Between Oregon and Dixon is where the Plum River Fault Zone gives way to the Sandwich Fault Zone.

Illinois Route 2 (IL-2): Devil's Backbone

The town of Oregon is situated adjacent to the west side of the Rock River in Ogle County. It is a midsized Illinois town with about 3,700 people living in it. IL-2 and IL-64 are the major roads in the town. Oregon sits on top of stream sands deposited by the Rock River and older Wisconsin glacial outwash of the Henry Formation.

As you head south on IL-2 from Oregon you will start to notice large bluffs. These are mostly the Ordovician Saint Peter Sandstone. The most northern exposures of the Saint Peter occur at the Devil's Backbone, which is a high west-northwest trending ridge. Like most of the bluffs in the area, it is the Saint Peter, but this outcrop is different from the typical Saint Peter that is common in the area.

The outcrop is called the Gale Creek/IL-2 Section along the Devil's Backbone. Here, in the vegetated rout cut on the west side of IL-2, is where the fault plane of the Sandwich Fault Zone is exposed. The Saint Peter is usually friable or weak quartz sandstone that can be easily scratched out with a blunt metal tool. This outcrop is very different. Here the Saint Peter is extremely hard. It is actually a white quartzite (the metamorphic version of sandstone). This localized type of metamorphism is called contact metamorphism and it is restricted to the northern face of Devil's Backbone. The outcrop is also marked by offset fractures on the east side. These are small fault planes. The north face of the outcrop contains large angular quartzite breccia. Here you are looking directly at the plane of the Sandwich Fault Zone. It is very rare, anywhere in the Midwest, that you get the opportunity to look at a fault plane head on. Usually they are expressed at the surface in cross section or obliquely.

GPS Location: 41.9850 -89.3376
Photograph is looking west showing the offset beds of smaller faults at the Gale Creek/IL-2 Section. Here the Saint Peter has been metamorphosed to quartzite from friction and pressure created during faulting.

GPS Location: 41.9850 -89.3376. Photograph is looking south at the fault plane of the Gale Creek/IL-2 Section. The Saint Peter is also metamorphosed here but the angular blocks visible are called breccia. The breccia formed as parts of the Saint Peter broke along the fault plane.

Illinois Route 2 (IL-2): Castle Rock State Park, Hard Rock Mound Section

On the west side of IL-2, about a mile south of Devil's Backbone is a small road cut with a big story. The rock here is dolostone with brown chert nodules. Overlying the Ancell Group of rocks in the area is the Platteville Group, which is a cherty dolostone, but these rocks are different. If you stand back on the east side of the road and look at them you will notice that the rocks sag in the middle. A closer look at the rock and you will see small carbonate crystals all of the same size. This is an indication that the rock has been recrystallized, destroying any small internal structures such as fossils and oolites. Faulting caused the dolostone to recrystallize in this outcrop just as it did at Devil's Backbone (previous page). These rocks are not Platteville. They are the older Cambrian Potosi Formation. A well located less than a quarter of a mile east of the outcrop (on Brooks Island) encountered the green sands of the Franconia Formation at less than 300 feet below the surface.

*GPS Location: 41.9766 -89.3577
Photograph is looking west at the sag in the Cambrian Potosi Formation. This structure was formed during several faulting events along the Sandwich Fault Zone. It is a combination of a thrust fault bounded by strike-slip faults.*

This location shows all the three main type of faults. It was originally lifted up as a horst during reverse faulting (compressed from north to south), then thrust up to the east in strike-slip faulting (compressed from east to west), then dropped down in a graben in normal faulting (extended from north to south). This area of the Sandwich Fault Zone was very active at different times and at this end of the fault there are many complicated relationships.

Close-up of the recrystallized dolostone in the Potosi Formation showing the small rhombohedral crystals.

Illinois Route 2 (IL-2): South Grand Detour Section

Further south along IL-2, just south of the Rock River and the town of Grand Detour about ¾ of a mile, is a low lying outcrop on the east side of the road. This outcrop is the South Grand Detour Section. This little outcrop has a lot to show. It is one of a small handful of accessible outcrops showing the Saint Peter and Glenwood Formation contact. The top of the Saint Peter here has horizontal red and green streaks within it. The red is iron stained sandstone and the green is glauconite and lesser pyrite. Glauconite and pyrite are rare in the Saint Peter overall, but they are common in the area of the Sandwich Fault Zone.

GPS Location: 41.8804 -89.4284 Outcrop looking east-northeast. The prominent ledge in the right half of the photo is the sandy carbonates of the Glenwood Formation. The white ledge at backpack level is the quartz sandstone of the Saint Peter Formation.

The red at this outcrop appears to be primary, which means it was deposited around the time of the Saint Peter itself. It is possible during the formation of the Sandwich Fault Zone, that heated water from deep below altered the mineralogy of the sand in what is called hydrothermal activity. It is also possible that the Saint Peter was exposed to the surface before the Glenwood Formation was deposited. In some places throughout the area, the Saint Peter-Glenwood contact is a minor angular unconformity. Red is a common color in sedimentary rocks that have been exposed to the surface for long periods of time, leaving the surface to oxidize.

Close-up of the green and red laminations at the top of the Saint Peter, looking east.

As you continue south to Dixon Illinois towards I-88, you will notice several high cliffs on private property. These are the dolostones of the Platteville Group. The Platteville Group is Ordovician in age but it is younger than the Saint Peter and Glenwood Formations. The Platteville Group also marks a change in the environment. The seas were deeper than in the time of the Saint Peter and favored carbonate deposition.

US-20: Rockford to East Dubuque

US-20 from Rockford to East Dubuque, takes you across the transition from the glacial covered Rock Hills Country to the unglaciated Driftless Area. As you head west on US-20 for the nearly 90 mile journey, you slowly come off the Wisconsin Arch but are north of any major faults. As a result the bedrock generally gets younger from east to west. You start out in the Middle Ordovician and slowly journey up section into the Upper Silurian rocks.

U.S. Route 20 (US-20): Freeport Sections

As you travel along US-20 from the Rock River Hills Country, to the Driftless Section you encounter isolated outcrops of bedrock above the highway in road cuts. In the vicinity of Freeport, in Stephenson County, these road cuts belong to the Middle Ordovician rocks of the Galena Group. The rocks are yellowish in color and are dolostone. They were deposited on a shallow marine shelf during a time of geologic stability in the area.

GPS Location: 42.3128 -89.5650
Photograph shows the typical of the Galena Group along US-20 near the junction with IL-75, looking northwest. Notice the uniform flat bedding of the rock.

U.S. Route 20 (US-20): Apple River Canyon State Park

Apple River Canyon State Park is located about five miles north off of US-20 at East Canyon Road, in JoDaviess County. These rocks are also part of the Galena Group, just like at Freeport. Unlike at Freeport, Apple River Canyon State Park is deep within the Driftless Section and was never glaciated, but it was near the front of the Illinois Episode ice sheet. For a time around 150,000 years ago, a slack water lake occupied the area. Once it drained to the southwest, the Apple River changed course for a narrow straight stretch. This event gave the Apple River its chance to carve steep cliffs in the dolostone rock thus exposing bedrock usually hidden by glacial drift elsewhere in Illinois.

One such exposed structure shows a shallow dipping monocline in the Dunleith Formation of the Galena Group. Other structures include large fractures and small caves dissolved out of the rock by groundwater. In other parts of the Park, the alluvium left by the Apple River contains unusually large rounded cobbles of the local rock. This is how most small streams in Illinois looked before the glaciers covered most of it.

GPS Location: 42.4447 -90.0574
Here the Platteville Group changes from flat bedding and dips south along a subtle unnamed monocline, looking southwest.

U.S. Route 20 (US-20): Woodbine Section

Due south of Apple River Canyon and slightly west on US-20 is an outcrop ascending a hill on the south side of the road. This is one of the few good outcrops of the Upper Ordovician Maquoketa Group. The outcrop is several hundred feet long and grades upwards from the hard dolostone of the Fort Atkinson Formation to the blue and green shale of the Brainard Formation. Here the Brainard Formation contains thin beds of argillaceous dolostone that are rich in fossils. The fossils are easily picked from the softer shale rock. The shale contains abundant fossils of bryozoans and brachiopods.

GPS Location: 42.3475 -90.1027 This photograph shows the dolostone and shale of the Maquoketa Group. The Fort Atkinson Formation is in the distance at center left. The Brainard Formation is at the center right. Photo is looking east along US-20.

This is a close-up of the Brainard Formation looking south. It is mostly shale with thin beds of protruding dolostone.

U.S. Route 20 (US-20): Terrapin Ridge Outcrops

West of the Woodbine section and just east of the small town of Elizabeth, stands a prominent ridge called Terrapin Ridge. This ridge is higher than the surrounding area because it is capped by Silurian dolostone, most of the surrounding area is low lying hills of Maquoketa Group. Since the Maquoketa is mostly weak shale and soft carbonates, it erodes down to low rolling hills. The Silurian is hard carbonate rock that is more resistant to erosion. The Sweeny Formation is exposed at the top of the Ridge and consists of a highly porous, thin bedded dolostone. On the flanks of the ridge the Blanding Formation is exposed with its white chert beds.

*GPS Location: 42.3151 -90.1973
Here are the dolostone of the Sweeny Formation on top of Terrapin Ridge, looking northwest on US-20.*

*GPS Location: 42.3158 -90.2040
This photograph is on the west edge of Terrapin Ridge and the very cherty Blanding Formation is exposed, looking north-northeast.*

U.S. Route 20 (US-20): Town of Scales Mound Section

Three miles south of the Wisconsin boarder and north about 10 miles from Elizabeth and half a mile north of Stagecoach Road, sits the small but geologically significant town of Scales Mound. 0.27 miles northwest of the junction of North Avenue and the railroad tracks is the type section of the Scales Formation. These are active railroad tracks, so proceed with caution! All formally named geologic units have to have a type section. This is so other geologists can use it for reference. Over time type sections may become destroyed for whatever reason, usually urban expansion is a main cause. If a type section becomes destroyed a primary reference section can be designated.

On both sides of the railroad cut is about three feet of the Dubuque Formation at the very top of the Galena Group. Here an unconformity is present separating the Dubuque Formation from the softer dark brown shale of the Scales Formation. The Scales Formation is the basal formation of the Maquoketa Group and is present throughout all of Illinois, but is usually buried under younger rocks. Here the rocks sit at near the same elevation as at Terrapin Ridge, but the rocks are older. As you go north from US-20 to the Wisconsin border, you cross a series of small anticlines and synclines. The net effect is older rocks are exposed to the north.

GPS Location: 42.4791 -90.2563
The basal vertical slops near track level are the Dubuque Formation. The vegetated slopes above are the Scales Formation. The photograph faces west.

A close-up of the weak brown shale of the Scales Formation, looking north, with 12 inch hammer for scale.

U.S. Route 20 (US-20): Town of Galena, East Entrance Outcrop

As you continue west on US-20 you will notice that the topography changes from rolling hills to high hills and cliffs. One town set back in theses cliffs is the town of Galena. Galena is known for its wine and the home of Ulysses S. Grant, but there is a lot of history in the rocks as well.

GPS Location: 42.4098 -90.4249 Galena Road Cut Section is a narrow pass in the Wise Lake Formation (upper 33 feet) and the Dunleith Formation (bottom 13 feet) on both sides of US-20, looking east.

Galena is the town where the Ordovician Galena Group of rocks gets its name. Galena is also a metallic, heavy mineral that was mined in the area for lead. This area is known to miners as the Zinc-Lead District. Most of the lead mined from the area is found in the Wise Lake and upper Dunleith Formations, as where most of the zinc is found in the lower Dunleith down through the Grand Detour Formations.

CHART OF THE ORDOVICIAN GALENA AND PLATTEVILLE GROUPS IN NORTHERN ILLINOIS

GROUP	FORMATION	SELECTED MEMBER	DOMINANT LITHOLOGY	* THICKNESS	MINERALS
Galena (Trenton)	Dubuque		argillaceous dolostone	0-40	None
Galena (Trenton)	Wise Lake		massive dolostone	30-80	Lead
Galena (Trenton)	Dunleith		dolostone, limestone with bentonite	50-135	Lead, Zinc
Galena (Trenton)	Decorah	Guttenberg	argillaceous dolostone with bentonite	0-30	Zinc, Petroleum
Galena (Trenton)	Decorah	Spechts Ferry	shale with bentonite	0-40	Zinc
Platteville	Quimbys Mill		dolostone, limestone with chert	10-30	Zinc
Platteville	Nachusa		cherty dolostone	15-40	Zinc
Platteville	Grand Detour		argillaceous dolostone with chert	45-180	Zinc
Platteville	Mifflin		argillaceous dolostone, limestone with breccia	15-30	None
Platteville	Pecatonica		crystalline to argillaceous dolostone	35-50	None

*Thickness is in feet.
The Galena Group is equivalent to the Trenton Group further south.
The Dubuque Formation is absent in Northeastern Illinois.
The Guttenberg Member contains localized petroleum.

Large and beautiful crystals have come from the mines and quarries of this area when lead and zinc were actively mined. Today only aggregate is mined in the quarries.

Minerals from a closed mine near Galena. The gold colored minerals are pyrite. The white colored minerals are calcite. The metallic gray colored mineral on the right is Galena. (Steven Baumann's personal collection)

U.S. Route 20 (US-20): Menominee Section

West of Galena, as you head towards East Dubuque, you will notice an outcrop on the north side of US-20, where the beds appear to dip slightly west near the Dunleith-Wise Lake Formation contact. This structure is just west of the Menominee River, and the east half of the structure has been eroded. It does appear to be the flank of a small syncline structure. It may be a small monocline. Without the west half of the structure it is hard to tell. Such minor structures are common in the area and most are undocumented.

The outcrop contains many white calcite veins, brown and red shale, along with brown chert nodules. Calcite will often fill small fractures as groundwater moves through the carbonate rock, dissolving out minerals and redepositing them in cracks and fractures. This mechanism also deposited most of the minerals in the area. The brown chert is primary, and was deposited when the carbonate rock was deposited on a seafloor during the Ordovician. The red shale represents a small influx of clastic sediments. The contact between the Dunleith and Wise Lake Formations is gradational. This means that there was no break in deposition (no unconformity). There is a zone of several feet where the transition between the two formations is represented by rocks of both formations.

GPS Location: 42.4654 -90.5899 North side of US-20. Notice the brown chert near the base and the nearly horizontal white calcite veins, looking east.

U.S. Route 20 (US-20): JFK Road Section

The frontage road off US-20 offers many outcrops for study between Menominee and the Mississippi River. This small outcrop on the northeast side of US-20 Frontage Road and JFK Road, is one of the few that are easily accessible and clear year round. Here you want to park on the west side of JFK Road. The outcrop is only about 10 to 15 feet tall but it is a good reference for the Dunleith Formation. The Dunleith forms most of the cliffs along US-20, and the exposures along US-20 are the type section of the Dunleith Formation. The large brown to gray chert nodules, within the formation are a key identifying marker. The Dunleith Formation is also very pure and contains almost no shale, unlike the formations above and below.

GPS Location: 42.4729 -90.6076
Brown chert in the Dunleith Formation, looking east.

If you look at the top of the outcrop you will see many loose and large angular boulders, in a three to four foot thick unit, immediately under the thin black soil. This type of large jumbled rocks on top of a flat surface is hard to explain. There is no evidence that these boulders were moved by ice (such as striations) or water due to their large size and angularity. Plus, this is the Driftless Section, so glacial transport isn't a viable possibility. There also is no evidence of faulting, such as slickensides on the boulder or rock face. The boulders were most likely deposited during a rockslide millions of years ago before the Quaternary, when the top of this outcrop was actually the bottom of a gully. An actual age for this ancient rockslide cannot be determined. It likely occurred before the establishment of the Mississippi River in the area. Perhaps 20 to 40 million years ago.

GPS Location: 42.4729 -90.6076
Outcrop of the Dunleith Formation with a large boulder field in the top half of the outcrop, looking southeast, backpack for scale.

U.S. Route 20 (US-20): Sections around East Dubuque

There are many outcrops in the East Dubuque area, too many to mention them all here. The outcrops are mostly carved out by human activity, to accommodate buildings. However, some are natural cliff faces. The natural faces tend to be a darker brown and much more weathered than the manmade cuts in the rock.

GPS Location: 42.4943 -90.6458
Outcrop of the Upper Galena Group in downtown East Dubuque on Sinsinawa Avenue, looking north.

GPS Location: 42.4840 -90.6320
Type section of the Dunleith Formation, along US-20 just east of Timmermans Supper Club, looking north.

GPS Location: 42.4747 -90.6170
Outcrop of the Upper Galena Group on US-20 and Tomahawk Road, behind a car shop, looking north.

GPS Location: 42.4737 -90.6124
Outcrop of the Upper Galena Group (Wise Lake or Dunleith Formation) in East Dubuque along US-20 in an area of new construction (during 2012), looking north.

I-39: Rockford to I-88

Interstate 39 (I-39): Rotary Road Section

On the west side (southbound side) of I-39 near the Rotary Road overpass, 0.62 miles south of the Kishwaukee River, is a small outcrop that looks out of place from the surrounding carbonates of the Galena Group. The rock is either the Wise Lake or Dubuque Formations (the two become hard to distinguish in the area). There is a small circular shape in the rock resembling an upside down bowl (dome) that has been cut in half. It is surrounded by flat rock. The structure is natural and the only known one in the area. If it was complete it would be almost circular and would be called a dome. Although not faulted at the surface, the rock is cracked as if it has been punched up from below. There are several things that can cause such a structure. One way is by fluids (usually salty) rising up from below in a salt dome. Another way is from regional stresses acting from different horizontal angles that were localized at this road cut. A third possibly is that the structure is a small surface reflection of a deeper fault, or perhaps a fourth unknown cause. There isn't enough data in the area to draw any solid conclusions on the origin of the structure.

*GPS Location: 42.1742 -89.0240
Outcrop of the Upper Galena Group (Wise Lake or Dunleith Formation) showing the small half dome structure, looking west, yellow seven inch tall notebook for scale.*

Interstate 39 (I-39): I-39 Sections

South of Rotary road until the exit for Illinois Route 64 (IL-64), several outcrops, generally less than 15 feet high, can be seen. These rocks belong to the Wise Lake Formation. The drive from Rotary Road may look flat but you actually descend about 40 feet in elevation as you head south towards IL-64. As you look at the outcrops you will notice that many have a prominent horizontal gray bed surrounded by more light brown colors above and below. This gray bed is a good local marker horizon for drillers and helps identify the Wise Lake in the subsurface.

*GPS Locations: Between 42.1829 -89.0238 and 42.0113 -89.0207
Outcrop of the Upper Wise Lake Formation with prominent gray bed, looking west.*

IL-84: The IL-84 and US-20 Junction to Savanna

IL-84 begins at the Wisconsin border and hugs the Mississippi River for most of its length. As it climbs and descends the bluffs along the Mississippi, it takes you on a journey from the Silurian to the Ordovician. The best exposures occur within the Mississippi Palisades Park area. The route lies entirely within the Driftless Area, although the area was never covered by ice, there are vast deposits of loess in the area.

Illinois Route 84 (IL-84): Mississippi Palisades State Park, Outcrop of the Galena Group

On the northwest corner of Mississippi Palisades State Park is a 10 foot outcrop on the east side of IL-84. This outcrop is the part of the Silurian, which overlies the Ordovician Maquoketa Group. These rocks are mostly dolostone. At this outcrop a spongy looking substance that coats the surface. This is tufa. Although it looks spongy, it is extremely hard. It forms on the surface of carbonate rock outcrops where groundwater continuously percolates from the rock. The groundwater partially dissolves the rock as it moves through it and redeposits calcium carbonate (limestone) on the surface.

GPS Location: 42.1556 -90.1796
Close-up of the tan colored tufa on the surface. Although it is spongy in appearance it is hard. U.S. dollar coin is for scale, looking east.

Illinois Route 84 (IL-84): Mississippi Palisades State Park, Ozzie's Point

Ozzie's Point is an excellent place to look out over the Mississippi River Valley. The rocks here are also Silurian and are capped by several feet of loess. As the Mississippi River erodes its channel laterally, this area is actively being eroded in what is called a cut bank.

GPS Location: 42.14062 -90.1650
Photograph is looking out west over the Mississippi River Valley from the edge of Ozzie's Point.

Also visible here is evidence for the erosive power of water. Carbonate rocks commonly form vertical fractures which rain water will percolate down into. As the rainwater moves through the rock, it dissolves out the dolomite and calcite. This turns the fractures into crevices and caverns. Eventually enough carbonate dissolves out, leaving the rock standing by itself in what is called a stock. As erosion continues stocks eventually collapse onto the surface in a rock fall and can block roads.

Here the erosive power of water has separated the rock forming a crevice at Ozzie's point, looking north.

Illinois Route 84 (IL-84): Mississippi Palisades State Park, Sunset Trail Section

Sunset Trail begins in the parking lot at Mississippi Palisades State Park. As you ascend the staircase of the trail you notice some carbonate rocks to your left. There are several things going on in at this outcrop. From trail level up to about three feet are dolostones with diverse but isolated fossils. Right above this is a three to four inch bed or layer of chert. For another foot above the chert is a very bumpy surface of dolostone with small iron concretions at the top. The iron concretions mark a minor angular unconformity with the rocks above. This type of unconformity is common within the Ordovician and Silurian carbonates and is called a diastem. Diastems are minor unconformities recognized by mineralized beds (hardground) below the unconformity. Hardgrounds are formed when deposition stops, usually only for no more than a few thousand years, and the seafloor gets either oxidized or pitted. In this case, oxidation occurred effectively "rusting" the sea bottom leaving behind the iron concretions. The diastem is probably at the contact between the Tete des Morts and Blanding Formations.

*GPS Location: 42.1278 -90.1556
Photograph is looking out west.
Person's hand is on the diastem unconformity. The person's elbow is at the chert bed.*

Illinois Route 84 (IL-84): Mississippi Palisades State Park, High-Face Section

Further south along IL-84 and just north of Savanna is another outcrop on the east side of the road in the south end of Mississippi Palisades State Park. The Silurian rocks here look fresher than they do to the north. The outcrop here was along a now abandoned railroad pass. The tracks were removed decades ago.

The Silurian rocks exposed at this long continuous outcrop are of the basal part of the Silurian and belong to the Blanding Formation (basal white cherty 30 feet), the Sweeny Formation (55 feet above the Blanding), and the Marcus and Racine Formations (inaccessible upper 80 feet). This is the type section of the Sweeny Formation, which is a gray to pinkish gray, thin to medium wavy bedded dolostone with shale partings.

Here the elevation is 30 feet higher than at Sunset Trail but we are lower in stratigraphic succession. This means the rocks here are about the same age or slightly older than at Sunset Trail. If the rocks were perfectly level, that wouldn't be the case, they would be younger. If you look at the rocks carefully in this outcrop they appear to dip slightly north 3° to 11°. This is because the outcrop just on the north side of the Plum River Fault Zone. The Plum River Fault Zone trends east-west and enters Illinois from Iowa almost right at the toll bridge into Iowa at Savanna. The rocks south of the fault are older rocks of the Ordovician Maquoketa and Galena Groups, indicating that the Silurian rocks were dropped down (relative to the Ordovician rocks) a couple of hundred feet during faulting.

GPS Location: 42.1086 -90.1586 Photograph shows the smooth rock at the base, with the very cherty (light colored bands) Blanding Formation and the non-cherty Sweeny Formation above, looking south.

Northeastern Illinois

Chicago Lake Plain

The Chicago Lake Plain is the smallest of the physiographic sections on land, although it does extend beneath all of Lake Michigan. It is only present in Cook County and a very small sliver of the southeast corner of Lake County Illinois and a small portion of Lake County Indiana. Although the area is small, it is perhaps the most dynamic of the physiographic sections.

The Chicago Lake Plain is named because at one point in the relatively recent past, it served as the floor of what was called Lake Chicago (a name given for Lake Michigan at its peak depth about 14,000 years ago). After the glacier that resided in the Lake Michigan Basin began to retreat, Lake Chicago formed at its front. The average surface elevation of Lake Michigan today is about 575 to 580 feet above mean sea level, at its height; it was 640 feet above mean sea level. That's almost 40 feet higher than the surface of Lake Superior today. We call the high stage Lake Chicago because at the time there was no true Lake Michigan at the time, since a glacier still occupied most of the Lake Michigan Basin. There are ancient sand beaches and shorelines now covered by roads and buildings. The majority of the area contains lake silt and clay beneath the modern urban fill.

Although nature has sculpted the landscape of the Chicago Lake Plain with ice and water, man has done so with bull dozers and concrete. The shape of the Chicago Lake Plain has also been shaped so much by human activity over the past 200 years that the original shape and configuration of the shoreline is all but forgotten.

The Cook County Faults are the dominant bedrock structure and are mostly lay under Chicago in the Silurian rock. The offsets range from a few feet to dozens of feet. Their origin is not known. They were discovered during the construction of the "Tunnel and Reservoir Plan" (TARP) project. The TARP Project is far from being completed. It has been under construction since 1975. TARP relies on deep drilled tunnels connecting to manmade basins to control flood water. Thornton Quarry was tied into the system in 2015 and is now called the "Thornton Composite Reservoir" and holds about 7.9-billion gallons of water. The last full tour of preflood quarry was in 2013, but limited tours are still being conducted.

Wheaton Morainal Country

As you drive through DuPage, Lake, most of McHenry, and the eastern part of Will Counties you notice that it isn't as flat as driving in McLean or Champaign County. This area is the Wheaton Morainal Country. Its topography is so much different because this is where the glaciers made their last stand 18,000 years ago before permanently retreating into the Michigan Basin. The hills represent many end moraines with many names. Most of them belong to the Wadsworth Formation. Most of the surficial deposits are glacial diamicton with peat and lake deposits.

The hills that were left behind by the glaciers are also dotted with many small lakes called kettle lakes. These are short lived lakes that if left untouched eventually shrink to bogs then dry up completely. They are a major source of peat in Illinois. It isn't all end moraines and lakes. Many of the rivers that flow in more north-south directions were once ice marginal rivers, such as the Fox River. They were at the front of the glaciers and are filled with sand and gravel deposits, which are another resource of Illinois.

Since the topography of the Wheaton Morainal County is so young, the natural drainage of the area is poor. This leads to the formation of many swampy areas, which are still present in the area and are a major concern to developers. They are not good places to build structures on because the soils tend to be soft and the clays are expandable. This can cause uneven settlement in buildings and form foundation cracks.

The Des Plaines Disturbance is a six mile diameter structure that lies in the area just off the northeast corner of DuPage County. Next time you are driving over I-90 near O'Hare (the southern tip of the impact crater) or passing along I-294 between Lake Road and Touhy Avenue (center of the north-south axis of the impact crater), you are driving over a place where a 1,000 foot diameter object struck the Earth millions of years ago (Volume I, p.49).

Kankakee and Bloomington Ridged Plains

The Kankakee and Bloomington Ridge Plains are present in both Northeastern and Central Illinois. Since they reside in both parts, they will be discussed in detail under Central Illinois (p. 62). For the purpose of this book, the southern boundary of Northeastern Illinois follows the southern border of Kankakee County and stretches from the Indiana border west to the Galesburg Plain. This is an arbitrary boundary selected for this book in an effort to define areas in a recognizable fashion and to avoid making either Northeastern or Central Illinois too large.

Guide for Northeastern Illinois

IL-137 and US-12: Lake Michigan to McHenry

Lake Michigan near the Wisconsin border is an area of formed by relatively modern processes. The sand dunes that trend the lake formed in the last 10,000 years on top of glacial diamicton and outwash that is only slightly older. As you head west, you enter topography typical of the Wheaton Morainal Country. These are the youngest glacial deposits in the state. They were deposited between 12,000 and 15,000 years ago. The result is a very young topography with poor drainage. The land is covered by large yet shallow glacial lakes and bogs, such as the Chain 'O Lakes, sit in the low areas between glacial end moraines. The pattern of roads along this 30 mile stretch reflect the geography as they twist and turn to avoid thick peat deposits of bogs and meander their way around end moraines.

Illinois Route 137 (IL-137): Illinois Beach State Park

The town of Zion is at about the halfway point along the long and narrow strip of Illinois Beach State Park, which stretches almost continuously, from Greenwood Avenue in Waukegan north to the Illinois-Wisconsin Boarder. Although the park is almost 8 miles long (north to south) it is only about one mile wide. Its western border is the Lake Border Moraine and its eastern border is Lake Michigan.

If you look at an aerial photograph you will notice about 20 or so linear features that parallel the shore of Lake Michigan. These are sand dunes that were former beach fronts. If you took a cross section along Wadsworth Road from Holding Ridge Avenue to the Lake, you would notice that the Henry Formation is the dominant surface unit with areas of Grayslake Peat. These overlay the glacial till of the Wadsworth Formation. The Henry Formation is locally beach sands and dune deposits of the Parkland and Ravinia facies. The Grayslake Peat formed where water pooled between ridges. The Equality Formation is also present in discontinuous areas. The Equality Formation is lake bottom silts and clays. It forms the majority of the Lake Michigan bottom. Wadsworth Road enters the Park at about GPS 42.4300 -87.8100

Illinois Beach State Park is also home to the Dead River. The Dead River is a small stream that twists and bends (meanders) its way through the park before emptying into Lake Michigan. There is a 1.8 mile hiking loop around the River. It is currently the only river in Illinois that still flows into Lake Michigan. Its mouth is a great place to observe both gravels deposited.

*GPS Location: 42.4088 -87.8036
Photograph is looking east, along the north bank of the Dead River. Near the backpack are unconsolidated sands and gravels. Lake Michigan is in the background.*

U.S. Route 12 (US-12) and Sullivan Road: Volo Bog

Off of US-12, at the Lake-McHenry County border sits Volo Bog. The bog was originally a larger series of kettle lakes that have become slowly filled with sediment and peat over the past 13,000 years. Kettle lakes are glacier remnants of ice blocks trapped under the sediment left by a retreating glacier. When the ice block melts it leaves a roughly circular shallow depression filled with water. It was originally much larger but is now restricted to a small area and will probably completely fill in within the next few centuries without interference from people.

*GPS Location: 42.3520 -88.1870
Photograph is looking south-southwest from a small hill on the north side of Volo Bog.*

IL-31, IL-25, and IL-62: McHenry to Aurora

From McHenry you can turn south along the state highways to Aurora for about 40 miles, roughly following the Fox River. Here you leave the youthful topography of the Wheaton Morainal Country and enter the slightly older Bloomington Ridged Plain. The Fox River, like the Rock River, cuts through the glacial deposits and exposes the mostly Middle Silurian to Upper Ordovician bedrock. As you drive south, you pass over the Kankakee Arch. The Kankakee Arch along with the Wisconsin Arch are subtle yet significant structures since they separate the Illinois and Michigan Basins.

Illinois Route 31 (IL-31): Moraine Hills State Park

Moraine Hills State Park sits east of IL-31 and the Fox River along River Road. Like Volo Bog it is a series of kettle lakes. Here, the lakes are deeper and are taking much longer to fill with sediment and peat. The name is somewhat misleading. There are no end moraines in the Park. The hills seen are caused by different features. The underlying material is outwash sand and gravel from the Henry Formation. The Fox River served as an ice marginal outlet when the glaciers occupied the area east of Moraine Hills State Park. The nearest end moraine is the Fox Lake Moraine, which is about one mile east of the park.

So what are all of the hills made of? Most are kames. Kames form either beneath or at the front of glaciers as bottom ice melts and the heavier particles remain (sand and gravel) and the lighter ones (silt and clay) get carried away.

*GPS Location: 42.3229 -88.2250
Photograph is looking southwest from a hiking trail. It shows a small kame in front of Defiance Lake in Moraine Hills State Park.*

Illinois Route 31 (IL-31): Klasen Road Quarry

Further south on IL-31, less than 3 miles from the Kane County boarder, are many shallow quarries. These Quarries are almost all in the outwash deposits of the Henry Formation and are within 3 miles of the Fox River. One such quarry is visible from Klasen Road. The quarry is private property so don't trespass. You won't need to. From Klasen Road, as you look north you will see that the floor is filled with water. This is probably near the contact with the Henry Formation (above) and the Tiskilwa Formation (quarry floor). Only the Henry is mined here for sand and gravel. There may be some glacial till of the younger Lemont Formation on top of and intertonguing with the Henry Formation.

*GPS Location: 42.1903 -88.2766
Photograph is looking north from a Klasen Road. The far wall is mostly Henry Formation. The slightly gray unit near the middle is likely the Lemont Formation.*

Illinois Route 31 (IL-31): Quarry Park Section

Along IL-31 in Batavia, sits a small park on the west side of the Fox River called Quarry Park. This is a small park with a manmade beach and pond. It was originally a stone quarry. On the west wall near the north side of the Park, is an outcrop, not of glacial deposits, but of Silurian carbonates. Silurian carbonates form most of the bedrock in Northeastern Illinois and are extensively quarried for aggregate and stone. This particular outcrop is of the Sugar Run Formation. The Sugar Run is a thin to medium bedded, greenish gray, argillaceous dolostone. It was extensively quarried in the 19th Century for building stone. Most stone buildings that still stand from Aurora to Joliet Illinois were quarried from the Sugar Run or the underlying Joliet Formation.

GPS Location: 41.8439 -88.3103
Photograph is looking west at the Sugar Run Formation.
The clipboard in the lower left is for scale.

Illinois Route 31 (IL-31): Mill Creek Section

About a mile south of Quarry Park is a stream that enters the Fox River from the northwest called Mill Creek. Mill Creek has an exceptionally flat and wide bottom for how shallow it is. It contained a larger volume of water when the Fox River transported melted glacial ice. Mill Creek flowed northwest then, opposite of today's southeast course.

If you stand in the creek where IL-31 crosses it you will see cliffs about 15 feet high. This is the Joliet Formation, specifically the Markgraf Member. The Markgraf is compositionally similar to the Sugar Run Formation, except it tends to be more yellowish gray in color, instead of the greenish gray hue of the typical of the Sugar Run. Here there are also small caves hollowed out of the rock along the south side of the creek. These were formed by carbonates being dissolved along fractures from rain and groundwater movement.

GPS Location: 41.8223 -88.3253 Photograph is looking west at the Markgraf Member where it meets the IL-31 overpass. The overpass was built around the yellow colored, layered, Markgraf Member of the Joliet Formation.

GPS Location: 41.8225 -88.3256 Photograph is looking southwest at the small caves in the Markgraf. Mill Creek flows from right (northwest) to left (southeast) in the photo.

Illinois Route 25 (IL-25): Flag Stone Quarry Section in Batavia

On the east side of the Fox River, the main road following it is IL-25. Almost due east of Quarry Park lies a shallow stone quarry called Flag Stone Quarry, which is owned by Fox Excavating. They quarry slabs of stone here, mostly for walkways and retaining walls. This is private property, so please respect it and ask for permission to access it or view it from the sidewalk.

This quarry was originally a small bedrock outcrop with about 10 feet of Henry Formation on top. The bedrock is Silurian and consists of the Joliet Formation. The two lower members of the Joliet Formation are exposed here. The wall on the east side is mostly the Markgraf Member. The quarry floor consists of the Brandon Bridge Member. Here the Brandon Bridge is an almost white, coarse crystalline, dolostone that is mottled red with green shale. It has been informally called "Holiday Stone" because of its red, green, and white color.

GPS Location: 41.8450 -88.3050
Photograph is looking east from IL-25. Notice the wall in the background. The sharp contact between the loose sand and gravel of the glacial Henry Formation on top and the bedded hard Silurian rocks below is visible (left center near the stacked blocks).

Illinois Route 62 (IL-62): Crystal Creek Section

About 2.5 miles southwest of the Klasen Road Quarry off of IL-62 is a small creek called Crystal Creek, which is nestled in between "Lake In The Hills" and a small business park. If you venture to see the stream cut, be careful Crystal Creek is deeper than it looks.

There is a stream cut that isn't very large but it is the best place to see the Tiskilwa Formation. The Tiskilwa is the oldest and most extensive, Wisconsin Episode glacial diamicton unit in Illinois. It is also distinct from the younger Wisconsin drifts and tills of the Wedron Group. Most diamicton formations of the Wedron Group are gray or light brown in color. The Tiskilwa has a distinct pinkish hue. It also contains more sand and typically smaller and more rounded clasts than the younger Wedron Group formations. It was deposited directly by the glaciers from the last ice age between 19,000 to 25,000 years ago. In contrast, the younger formations of the Wedron Group were deposited 14,500 to 18,000 years ago. We know the age of the glacial deposits pretty well because of organic debris within that has been carbon dated at numerous locations throughout Northeastern Illinois.

GPS Location: 42.1718 -88.3058
Photograph is looking southwest from Crystal Creek. Notice the pinkish hue in the exposure.

Close-up of the diamicton of the Tiskilwa Formation, looking southwest.

US-34: Aurora Area

Aurora is built surrounding the Fox River. It is the second largest city in Illinois. As a result, development has covered much of the geology along the roads. There is a large quarry near I-88 that cuts through the lower Silurian to the middle Ordovician. It is private land and they do not run tours. There are small exposures in the streams and creeks around the outskirts of Aurora and the surrounding area. The Fox River served as an outwash channel between the Minooka and Saint Charles End Moraines.

U.S. Route 34 (US-34): Waubonsie Creek Section

Waubonsie Creek is a small stream in Oswego Illinois. It enters the Fox River from the east. As you travel south along the roads bordering the Fox River, the exposed rocks get progressively older. The best place to see the geology is downtown, between Adams Street and the IL-25 overpass. You can park at the Oswego Public Library, the outcrop is behind it and there is a little trail. As this book is being revised, the old Adams Street is gone and the construction is not yet complete with the new parking to the west, so you can't park there anymore. The outcrop is on the south side of the Creek.

The bedrock that you see here is different from the Silurian or Ordovician carbonates seen thus far in Northeastern Illinois. If you notice the rocks near creek level have a distinctive green tint. There are softer (shale) interbedded with harder fossil rich purplish gray dolostone. These rocks are upper Ordovician. They are near the transition zone between the green shale of the Brainard Formation (above) and the hard dolostone of the Fort Atkinson Formation (below), similar to what is present at the Woodbine Section in Northwestern Illinois. Under the railroad tracks at creek level the rocks are loaded with fossil fragments. One fossil is unique and it is named for this outcrop. The little creature is called *Tentaculites oswegoensis*. This fossil is typically from half an inch to one inch long and looks like a tiny ribbed horn.

GPS Location: 41.6853 -88.3519
Photograph is looking east from under the railroad tracks. The ledges on the right, a few feet above creek level, are the Ordovician Brainard Formation. The library is the building at the top of the photo.

At a location only about 100 feet east of the railroad tracks is a 15 foot outcrop of the alternating green shale and dolostone. Near the top a concrete block (which fell to creek level in 2010) marked the contact between the Ordovician and Silurian. The contact is easy to pick out because there is a paleosol (or ancient soil) present on top of the shale here. It isn't a soil as we typically think of it; there were no land plants at the time. It is a yellow, weathered zone. This paleosol formed at the end of the Ordovician when glaciers thousands of miles south, lowered the global sea level by hundreds of feet. This was a short lived ice age that occurred at the end of the Ordovician thousands of miles from Illinois. When the glaciers melted the area again became flooded, and the Silurian carbonates were deposited.

GPS Location: 41.68555 -88.3517 Photograph is looking south at the Brainard Formation. The concrete block near the top marks the Silurian-Ordovician contact. The yellowish paleosol is visible just to the right of the block. The concrete block fell into the Creek in 2010 and in February 2020, the block was all but gone. This photo was taken in February 2008.

As you head east you will notice that yellowish brown carbonates begin to appear that contain no shale. These are actually the basal Silurian dolostone of the Wilhelmi Formation. There is some chert in the rocks as well. Here the Ordovician is exposed at the very bottom in a recessed part of the cliff.

I-90/I-94, I-80/I-294: Chicago Area

The City of Chicago is built upon mostly glacial diamicton and lake deposits. The building of the city over the past 185 years has altered the original landscape of beaches and swamps by adding fill to extend the city's eastern limits into Lake Michigan, altering the skyline with massive skyscrapers, changing the flow direction of the Chicago River, digging large sewage tunnels deep hundreds of feet below the city, and the creation of several large bedrock quarries that have mostly been abandoned and now serve as landfills. Even with all of this alteration, the city is a hub of urban geology. The Water Tower is constructed of Silurian dolostone from the Lemont area. 333 North Michigan Avenue is faced with Minnesota Precambrian granite on its lower floors. The Chicago Cultural Center is faced with Precambrian granite from the Stone Mountain area in Georgia. The Pittsfield Building contains ammonites from the Jurassic Limestones in the Alps. Epiphany Church at 201 South Ashland, is constructed from red Precambrian sandstone likely of the Jacobsville Group from Wisconsin, the Upper Peninsula of Michigan, or Ontario. Although far and in between, some exposures of bedrock are visible in the Chicagoland area.

Interstate 90/94 (I-90/94): Addison-Kimball Section

The City of Chicago is known for a lot of things, its buildings, pizza, baseball teams, lake front, and countless other treasures. What Chicago is not known for is its bedrock outcrops. That's because there is currently only one naturally occurring bedrock outcrop within the city limits of Chicago. You better look carefully because it takes less than two seconds to pass it on I-90/94, unless you're in rush hour traffic. It is in the Avondale neighborhood on the west side of I-90/94, just south of the Metra line overpass. It is only several feet high and sits at about 600 feet above mean sea level, or roughly 20 feet above Lake Michigan mean level.

GPS Location: 41.9458 -87.7184
Photograph is looking northwest at the low laying outcrop. Notice how its light color stands out from the surrounding area. I-90/94 is visible at the right.

Although tiny, it is a significant outcrop. It is one of the few places to easily see the reef rocks of the Racine Formation. The Racine Formation is the youngest, and thickest of the Silurian Formations in Northeastern Illinois. This is only the small tip of a much larger reef that extends underground. Its exact size isn't known. Where the Racine Formation hasn't been eroded by the glaciers it is typically over 300 feet thick.

At this outcrop along the highway you see the carbonate rocks typical of the reefs of the Racine Formation. You will notice that it is loaded with all kinds of fossils and white calcite crystals. The rock, although very hard appears to be porous or vuggy. Carbonate rocks often have small holes throughout called vesicles. If they are large enough, they are called vuggs. Sometimes, they become filled with minerals and become geodes. Although the Racine Formation isn't known to contain any geodes, it does typically have minerals such as calcite within it. Reef rock is very pure limestone or dolostone. It rarely contains any sand or shale.

Close-up of the outcrop, looking down and west at a large white calcite crystal.

Close-up of the outcrop, looking down and west. The honeycomb structure in the middle is a fossil as and the larger pits contain other fossils.

Interstate 80/294 (I-80/294): Thornton Quarry Section

The previous outcrop only showed a small tiny peak of an ancient Silurian reef. There is one place in Illinois where you can see an entire reef, minus what they have quarried out. As you enter Illinois from Indiana on I-80/94, about 3.5 miles from the boarder on I-80/294 you notice road cuts on both sides of the road inside hard very light gray rock. The beds or layers are dipping to the east here. You are on top of the flank of an ancient barrier reef, similar to the Great Barrier Reef of today, except that it is no longer surrounded by ocean and is 600 feet above mean sea level. Less than a mile west of these road cuts you all the sudden pass over a giant, deep quarry that at its narrowest is about 0.4 miles in diameter and over 100 feet deep. This is Thornton Quarry. The interstate is not suspended over the Quarry. The quarry was dug around it and the interstate actually sits on solid rock.

GPS Location: 41.5797 -87.6098 Photograph is looking north at the low the eastward dipping beds of the Racine Formation about 1,200 feet east of the quarry. The dip of the rocks increases from 7° to about 22° as you travel from the west to the Quarry. The seven inch yellow notebook is for scale.

Thornton is one of the larger reefs, almost a mile in diameter and several hundred feet thick. From rock cores in the area we know that the Joliet Formation sits underneath it. The Sugar Run Formation is absent, as it often is under large reefs. As you pass over the Quarry you will once again see road cuts with dipping beds. However, the beds dip west here. That's because you are on the opposite side of the reef. Reefs build slowly in layers, but not flat layers. The layers tend to form mounds. The structures are sometimes called bio-domes. Instead of being formed by tectonic activity bio-domes form from eons of living things growing, dying, and growing on top of previous generations.

As at the I-90/94 outcrop the rocks are loaded with fossils. The reef at Thornton was a community of living things for perhaps as long as 10 million years.

Close-up of the rock texture. Notice the fossils and vesicular nature of the pure carbonate rock, looking north.

Close-up of a weathered fossil. Notice the large vug left as this coral has weathered away, looking north.

I-80, I-55, and I-355 Corridor: Naperville to Lemont to Channahon

The 20 to 25 mile stretch from Naperville to Channahon skirts the southwestern border of the Wheaton Morainal Country and enters the northern tip of the Kankakee Plain. The route contains two major Wisconsin Episode glacial advances and is represented by the Lemont and Wadsworth Formations. Scattered throughout are Ordovician and Silurian bedrock exposures. In 2007, the opening of the I-355 extension to I-80 greatly increased access to the small towns throughout the area.

Interstate 55 (I-55): Shepley Road at DuPage River Section

Not too far off of I-55, just on the fringes of Channahon Illinois, is a small outcrop at the Shepley Road Bridge over the DuPage River. The GPS location is 41.4680 -88.2089. It isn't a very large outcrop but it deserves mentioning. The outcrop is on the south side of the DuPage River and is only about 6 feet high. At first glance it appears to resemble Silurian carbonates, but it is older. This is the Ordovician Fort Atkinson Formation of the Maquoketa Group. The Fort Atkinson is referred to informally by drillers as the "Middle Limestone" or the "Divine Limestone". This formation occurs at about the halfway point in the Maquoketa and is present throughout Southeastern Wisconsin, Northeastern, and Central Illinois. It is also present in Northwestern Illinois but its distribution is patchy. Where its distribution is patchy, it is often not separated out from the other formations in the Maquoketa Group. The Fort Atkinson is typically about 30 feet to 50 feet thick.

In the Channahon area the Fort Atkinson is a coarse crystalline dolostone that is often mottled red, white, and black. To the untrained eye it can resemble granite at first glance. It is definitely sedimentary in origin, as where granite is igneous in origin and chemically very different. This outcrop contains good sized crystals of dolomite, near the contact with the overlying Brainard Formation. It has been extensively quarried throughout the area. Here, it is only several hundred feet from the Sandwich Fault Zone. In the area, the Fort Atkinson often contains hydrocarbons that were concentrated within it along the Sandwich Fault Zone. Whether or not the hydrocarbons are of any economic value has not been determined.

Interstate 55 (I-55): Barbers Quarry Section

Significantly north of the last outcrop, in DuPage County, is a relatively large quarry that is somewhat hidden. It is Barbers Quarry and is also along the DuPage River within the town limits of Naperville, along Royce Road. Here you are near the crest of the Kankakee arch. The Quarry is not accessible to the public but is observable from the neighboring park (DuPage River Park), which is a reclaimed quarry.

At this quarry, the basal five feet of the Sugar Run Formation, the entire Joliet Formation (and all of its members), and the two upper members (The Plaines and Troutman Members) of the Kankakee Formation are exposed. The basal member of the Joliet Formation, the Brandon Bridge Member, is about 20 feet thick in the Quarry and is about 10 feet from the bottom of the Quarry. Here it is an impure thin bedded dolostone capped with several feet of green to black shale. It is also deeply red in color. The red color comes from iron and is rare in the Silurian carbonates of Illinois. The seas that deposited the Brandon Bridge were precipitating iron as well as carbonates. This could have been caused by a chemical change in the atmosphere, perhaps by increased oxygen levels or a brief exposure above sea level.

*GPS Location: 41.7162 -88.0886
Photograph is showing the deep red color of the Brandon Bridge Member in a detached block that was located outside of the quarry in 2005, 18 inch hammer for scale.*

Interstate 55 (I-55): Waterfall Glen Section

Waterfall Glen is a large, yet somewhat concealed park in DuPage County. The focus here will be on the part that contains Sawmill Creek south of Argonne National Laboratory. You can enter the Park at the bend in Cass Avenue and Bluff Road, and then follow the trails, heading south-southwest.

Saw Mill Creek is a devious little steam that enters the Des Plaines River from the North. It served as an outwash channel when the Wisconsin Episode glaciers retreated. It has subsequently cut down through 20 feet to 50 feet of glacial sediments down to the Silurian bedrock. There is another smaller stream that parallels Sawmill Creek a few hundred feet to the east. This little stream use to occupy part of the southern area of Sawmill Creek and the two were connected. Sawmill, being the larger stream "hijacked" the lower part of the smaller stream. The name for this act of nature is called "stream piracy". The smaller stream now flows directly into the Des Plaines River. The abandoned channel that use to connect the two can still be seen as gravelly deposits near the Des Plaines River.

Sawmill Creek having a narrow valley, has exposed two glacial diamicton units. The lower Yorkville Member of the Lemont Formation is separated from the upper Wadsworth Formation by a three foot to four foot thick layer of the Henry Formation. The Wadsworth Formation is eroded through much of area surrounding the Creek, and the sandy gravels of the Henry Formation often lay underfoot along the higher trails.

*Location: 41.7049 -87.9646
Photograph shows the Henry Formation separating the brown and gray diamicton of the Wadsworth Formation above from the gray diamicton of the Lemont Formation below, looking south. The blue ball cap is at the base of the Henry Formation. Sometime between 2013 and 2015 this outcrop has been covered during a slope stability project. Sadly, it is no longer exposed.*

Interstate 55 and Interstate 355 (I-55 & I-355): Keepataw Forest Preserve

Keepataw Forest Preserve is a hidden away area about one third of a mile west of I-355 on Bluff Road near Romeoville, Illinois. There is a small parking lot on the south side of Bluff Road. There are trails extending from the parking lot. If you stay on them, you will not get to see the important geology exposed here. The Silurian-Quaternary contact is exposed about 250 feet south of the parking area and is extensively exposed in the bluffs. Here the glacial outwash deposits of the recent Henry Formation sit on top of a very flat and smooth contact with the Silurian Racine Formation. The Racine Formation at Keepataw, is interreef fine grained dolostone interbedded with white chert. Below the white chert is the finer grained dolostone of the Sugar Run Formation.

If you look further south, towards the Des Plaines River, you will see conspicuous mounds about 10 feet high. These mounds are quarry waste piles of rock. In these piles, fossils of trilobites and other marine creatures have been found. At the Des Plaines River itself, the Joliet Formation is exposed. It contains far more fossils than the overlying Sugar Run Formation.

GPS Location: 41.6779 -88.0370 Photograph shows the more massive Sugar Run Formation below the more platy and cherty Racine Formation, looking north-west.

The Des Plaines River area was extensively quarried for the buff colored dolostone rock that is common in the faces of older buildings of the Chicagoland area. Most of the rock quarried was either from the Sugar Run or Joliet Formations. During the time it was quarried, in the mid to late 19th and early 20th Centuries, it was called the "Athens Marble". Athens was the name of the nearby town of Lemont until 1850.

Interstate 55 and Interstate 355 (I-55 & I-355): Totora Food Section

Lemont is located about 1.75 miles east of Keepataw, and is on the south side of the Des Plaines River. The town of Lemont is not only a historical town but also boasts some of the best prehistory in Illinois. Lemont is one of the few towns where you can observe rock cliffs cut back in order to accommodate buildings. It is akin to the cut cliffs of Northwestern Illinois along the Mississippi River.

One good exposure of bedrock is located in the back of the old Totora's Food Center, (now home to Englewood Construction, Inc.) at the southwest corner of Lockport and Main Streets. In the south wall is the interreef rock of the Sugar Run and Racine Formation, similar to at Keepataw. Here the contact between the two formations is about five feet above the surface of the parking lot. The two formations can be difficult to tell apart. The Racine generally is darker in color and contains white chert.

GPS Location: 41.6722 -88.0042 Photograph shows the contact between the Racine Formation (above the notebook) and the Sugar Run Formation (below the notebook), looking south-southeast.

Interstate 55 and Interstate 355 (I-55 & I-355): Lemont, I&M Canal Section

The contact of the Joliet Formation and Sugar Run Formation is visible at the I&M Canal Trail just north of downtown Lemont. The trail sits at about 585 feet above mean sea level. This is about 20 feet lower than at the Totora Outcrop. If you walk towards the edge of the trail at the canal and look down, you'll see several feet of exposed bedrock going down into the water. Here there is about two feet of light yellowish gray fine grained dolostone above a few feet of a yellowish white, coarse grained, porous dolostone. The fine grained dolostone is the Sugar Run Formation. The coarse grained, porous dolostone is the upper Romeo Member of the Joliet Formation.

GPS Location: 41.6763 -87.9979 Photograph shows the contact between the Sugar Run Formation (above the pencil) and the Joliet Formation (below the pencil), looking east.

Illinois Route 171 (IL-171): Dellwood Park

Another example of catastrophic flooding occurred at Dellwood Park in South Lockport. A small stream flows through Dellwood Park called Fraction Run. It has sharp bends and high cliffs, yet is only a couple of feet deep. There is no way that such a small stream could have cut into the hard dolostone cliffs of the Racine and Sugar Run Formations, and create such a relatively wide flat valley. If you look at the cliffs you see that they appear to "bow out" in the middle. Most streams cut cliffs that are nearly vertical or steep sides that point away from the stream at the top and cut under the cliffs at water level. That isn't the case here. Here the valley is at its widest halfway up the cliffs. What could account for such an out of place feature?

If you look at the path of Fraction run, which flows almost due east to west, minus the tight bends, it looks like a normal creek at its eastern edge. As you head west into Dellwood Park, it drops in elevation about 50 feet and forms tight meanders until draining into the Des Plaines River Valley. It makes this dramatic change in less than a half a mile. If we look at local water well logs in the area, and construct a bedrock topographic map, we find something interesting. At about where Fraction Run makes its dramatic transition, we see a buried bedrock valley extending northwest towards downtown South Lockport. This northwest trending subsurface valley was the original valley occupied by Fraction Run before it became blocked by the glaciers of the last ice age. When the original path of Fraction Run became blocked, it sought a new route. This route would be the west route that it currently takes.

A redirection, does not explain the cliff cuts, just the path Fraction Run took. Fraction Run was originally 30 feet higher in elevation until the Kankakee Torrent released its flood waters. The Kankakee Torrent deepened the Des Plaines River Valley just west of South Lockport. As a result, all of the other tributaries lowered their valleys in response, and some are still doing it to this day. When Fraction Run was forced to quickly deepen its valley, it had to follow the fractures of the rock. This created the tight meanders. During the Kankakee Torrent, a much larger volume of water fed Fraction Run. This caused the cliffs to bow out when Fraction Run was significantly deeper, with much more water moving very quickly through it.

If you ever get to walk along the bottom of the Fraction Run Valley at Dellwood Park, don't forget to look up at the cliffs. Not only will you see the bow shape in the cliffs but also thin beds of light colored chert and green clay common in the interreef rock of the Racine Formation.

GPS Location: 41.5730 -88.0629 Photograph shows the concave cliff formed by the catastrophic outflow of water during the Kankakee Torrent. Maximum water level was about half way up the cliff, looking west downstream of Fraction Run.

Interstate 80 (I-80): I-80/US-30 (Exit 137) Section

Further west along I-80 at the Lincoln Highway (US-30) intersection is a small outcrop about 10 feet high on the southeast side of I-80 and along the westbound entrance ramp from Lincoln Highway on the northwest side. Here I-80 is at about 633 feet above mean sea level.

Most of the outcrop is the top member of the Joliet Formation and is called the Romeo Member. It closely resembles the reef rock of the Racine Formation and tends to be vesicular and rich with fossils. The main difference is that the Romeo isn't confined to small, local areas. It is regionally extensive and is present throughout northeastern and central Illinois. Along I-80 the beds dip about 3° southeastward. This is significantly different than the regional northeast trend in this part of Illinois. This is because you are on the south limb of the Kankakee Arch approaching the Sandwich Fault Zone.

GPS Location: 41.5164 -87.9953 Photograph is looking east-northeast at the slightly dipping beds of the Romeo Member.

GPS Location: 41.5182 -87.9947 Photograph is looking northwest from the entrance ramp to I-80 westbound. The Romeo Member is in contact with the overlying Sugar Run Formation. Blue ball cap is about one foot below the contact.

Also present at the top of the outcrop from about 640 to 643 feet above mean sea level is the argillaceous Sugar Run Formation, the same formation exposed at Quarry Park (p.26). Here, erosion has removed most of it, leaving only a cap remnant.

IL-83: Lemont to Alsip

The stretch of IL-83, where it runs west-east from Lemont to Alsip, generally follows the Cal Sag Canal. The Cal Sag was constructed as a shipping canal at the turn of the 20th Century. It cuts through the Lemont Formation and the Upper Silurian bedrock. Although manmade, the path of the canal does follow an old natural outwash channel that cuts through glacial end moraines.

Illinois Route 83 (IL-83): Swallow Cliffs

As you travel the strip of IL-83 (also known as Cal Sag Road), just northwest of Palos Park, you will notice a conspicuous mound on the south side of the road. This mound is called the Swallow Cliffs. It was the home of several toboggan ramps. The ramps were closed in 2005 and removed in 2007. Now the cliff is open for sledding in the winter.

This 60 foot cliff has a history before it became a winter recreation area in the 1930's. The cliffs were not cut by human hands, nor is it the debris piled up from the excavation and construction of the nearby Cal Sag Canal. It was produced by nature and is cut into a glacial end moraine. The Des Plaines River and several other streams served as outlet channels for the catastrophic flooding that occurred during the Kankakee Torrent about 14,000 years ago. As water from Ancient Lake Chicago searched for a place to drain, it quickly carved a channel through the local the end moraines. The flat area at the base of the cliff is covered by a thin layer of sand and muck overlying the Silurian bedrock.

GPS Location: 41.6808 -87.8625
Photograph shows the Swallow Cliffs formed by the catastrophic outflow of water during the Kankakee Torrent. Water flowed from the left (east) to the right (west) side of the photo.

US-52 Corridor: Joliet to I-39

The 50 mile stretch from Joliet west to I-39, transitions from the Kankakee Plain deep into the Bloomington Ridged Plain. The low hills of sand dunes give way to the Wisconsin Episode diamictons, with little exposure. Buried deep under the surface is the northeastern corner of the Illinois Basin. South and west of here the bedrock at the surface gradually becomes younger as the glacial deposits become gradually older.

County Road 32 (Co. 32): Sheridan-Fox River Section

Far away from any expressway, in the northeastern part of LaSalle County, lies the small town of Sheridan, about six miles due west of IL-71 from Newark. Sheridan itself sits on the glacial deposits of the Bloomington Ridged Plain. Although thin glacial deposits cover the area, there are no good glacial outcrops near Sheridan. Any good outcrops were long ago flattened and covered for either agricultural use or construction.

Just west of Sheridan on Co. 32, at the bridge over the Fox River, there is something much older exposed in the river bluffs. The eastern bluffs of the Fox River, display an orange, yellow, red, and white rock. This is one of the rare outcrops of the New Richmond Sandstone in Illinois. It is also the oldest sandstone visible in natural outcrops in Illinois.

The New Richmond strongly resembles the younger Saint Peter Sandstone (which we will visit later). There are several differences. The New Richmond is not as pure. It is mostly quartz sandstone, but it is often cemented with dolomite and it commonly contains impurities such as iron and feldspar grains. In the area, the New Richmond is used as a source of groundwater for local communities.

GPS Location: 41.5299 -88.6961
Photograph shows the multi-colored sandstone of the New Richmond Sandstone right above the water. This is looking southwest and down at the Fox River from the Co. 32 bridge.

IL-1: Near the Indiana Border

IL-1 generally is geologically bland due to vegetative and urban cover. There is one spot about 2.5 miles from the Indiana border and about 18 miles south of Crete that deserves a special mention. It is an old water filled quarry on private land along N15500E Road.

Illinois Route 1 (IL-1): Edgetown Quarry Section

Off of IL-1, in a slightly hidden away corner near the Indiana border, in Kankakee County, sits an old quarry filled with water. This is Edgetown Quarry. The Quarry is about 2.5 miles west of the Indiana border and about 4.8 miles southeast of Grant Park or the same distance east-northeast of Momence. It is on private land and you won't see it driving by because of a berm. It is about 40 feet deep. Quarrying stopped because there was too much groundwater was coming in and it could not be pumped out fast enough.

The exposed rocks are mostly very sandy, chalky, dolostones capped with a couple feet of very pure, very fossil rich dolostone, with a rubbly appearance. The bedding isn't flat. Beds pinch in and out and form small bio-domes. These are mini-reefs or reef mats and were formed by algae. The top dolostone contains trilobites and *Favosites*. The trilobite present is *Gravicalymene*. These are typical of late Silurian fossils, but they are also typical of Devonian fossils. Sandy reef mats are not known to be present in the Silurian in this part of Illinois, but are common in the Devonian. It is possible that these rocks are a remnant of a small patch of Devonian rocks that are equivalent to the Detroit River Formation in Northwestern Indiana. Here you are at about 630 feet above mean sea level, or roughly the same as at the I-80/US-30 Section.

GPS Location: 41.1917 -87.5748
Photograph is looking west. Notice the uneven, rubbly looking bedding.

IL-102 and IL-103: Wilmington to Kankakee

The trip along IL-102 and 103, between Wilmington and Kankakee, is a 19 mile stretch through Will and Kankakee Counties that straddles the Kankakee River. This is in the heart of the Kankakee Plain. The Ordovician, Silurian, and Pennsylvanian bedrock is covered by a thin mantle of glacial deposits. The glaciers left behind different features here than they did in the rest of Northeastern Illinois. This is a land of thick sand and gravel outwash covered with windblown sand dunes.

Illinois Route 102 & Illinois Route 113 (IL-102 & IL-103): Kankakee River State Park

One of the most majestic places to visit in northeastern Illinois has to be Kankakee River State Park. The park is sandwiched between IL-102 and IL-103 on its north and south ends. It sits about half way between I-55 and I-57 and is situated in both Kankakee and Will Counties. It is large enough to have amenities such as paved trails, brick lavatories, picnic areas, parks for children, and campgrounds, yet it is not as busy as the more famous Starved Rock State Park (which we will visit later). It gets its name from the Kankakee River which flows in a west-northwest direction through the middle of the park. It also has a tributary that enters it from the north called Rock Run. Rock Run formed in a similar manner as Fraction Run did (p.36-37). Rock Run also cuts though Silurian dolostone. The Silurian here is older, consisting of the lower Joliet Formation and the underlying Kankakee Formation. The Kankakee Formation is extensively exposed in the park. There is something unique exposed in the bluffs on the north side of the Kankakee River near the campgrounds. Here the Silurian-Ordovician contact is exposed. This isn't unique in and of itself. What is unique is the Ordovician unit exposed beneath the Kankakee Formation.

Kankakee River State Park is the only place in Illinois where the Neda Formation is exposed at the surface. The Neda is a purplish to red brown dolomitic shale with oolites. Oolites are small spherical balls less than one millimeter in diameter. They are usually made of white calcite, but not here. Here the oolites are made of iron minerals and are black to deep reddish brown in color. The oolites were originally deposited as calcite, some scattered white calcareous oolites still remain. At the end of the Ordovician the seas retreated due to an ice age thousands of miles away. This retreat of the sea exposed the upper Ordovician rocks to the air allowing them to become oxidized. Slowly the calcite was replaced with hematite and the original green color of the shale that contains the oolites was colored red and purple, with iron minerals (mostly hematite).

When the seas returned, the environment changed favoring the deposition of carbonate rocks instead of the clastic oolitic shale. Shale would not dominate deposits again in Illinois until the end of the Devonian Period. As we leave Kankakee River State Park we also leave behind the Silurian rocks that dominate the bedrock surface in Northern Illinois.

GPS Location: 41.2117 -88.0253
Photograph shows the Ordovician Neda Formation (below the black pencil), below the Silurian Kankakee Formation (above the black pencil), looking east.

GPS Location: 41.2023 -87.9813
Photograph shows the water cut cliff of Silurian dolostone on the shore of the Kankakee River, looking northeast.

US-6, IL-71, IL-178, and IL-351: Ottawa to Oglesby

The 13 mile journey from Ottawa to Oglesby crosses the most extensive and prominent structure in the state other than the Illinois Basin itself, the LaSalle Anticlinorium. As you head west on any of the east-west trending roads in the area, the bedrock goes from dipping gently east to dipping steeply west. The glacial cover in the area is generally thin and generally lacking near large streams and the Illinois River. The area has some of the largest parks in the state and is known for its vineyards.

U.S. Route 6 (US-6): Buffalo Rock State Park, N2803 Section

About 1.25 miles due south of US-6, and about 1.75 miles west of the outskirts of Ottawa, is a 30 to 40 foot outcrop off of the overpass at N2803rd Road. You can easily park off the overpass and descend down through the brush on the west side but be careful of the cliff drop. Once at the bottom you will see quartz sandstone that looks a lot like the New Richmond Sandstone at Co.32., but it is not. This is the Ordovician, Saint Peter Sandstone. This is a manmade cliff that extends from this location southwest for about one quarter of a mile until it reaches the railroad tracks. The cliff was mined for the pure quartz sand. It is now part of Buffalo Rock State Park.

If you examine the sandstone closely, you will notice that it is cross bedded in places and flat bedded in others. The Saint Peter was Locally deposited in a shallow marine environment. There are no fossils here. The Saint Peter being pure quartz, and almost all medium grained sandstone, did not allow for the preservation of fossils. They preserve better in limestone and shale. The Saint Peter is naturally white. Where surface and groundwater has moved through it near the surface, it takes on red, orange, and yellow colors. This coloration is caused by the deposition of iron that is present in the groundwater.

Sandstone wasn't the only thing quarried from these cliffs. The Colchester Coal is present at the top of the cliffs but is usually covered from view. The coal is relatively thick in the area (up to four feet) and was extensively mined in the early 20th Century. The coal is Pennsylvanian in age.

GPS Location: 41.3348 -88.8999 Photograph shows the multi-colored sandstone of the Saint Peter Formation. This is looking northwest at a small waterfall.

GPS Location: 41.3329 -88.9031 Photograph showing less weathered Saint Peter forming the cliffs. Notice the lighter color and flatter bedding. Here it is topped with gentle sloping Pennsylvanian shale, looking west.

U.S. Route 6 & Illinois Route 71 (US-6 & IL-71): Buffalo Rock State Park, Illinois River Section

Buffalo Rock State Park boasts many cliffs of the Saint Peter Sandstone. There are cliffs of Saint Peter along the entrance road and along the Illinois River. The Saint Peter is weakly consolidated sandstone. However, it holds together better than the weak shale of the overlying Pennsylvanian deposits. This creates some dramatic overlooks along the Illinois River.

Where the Saint Peter is present in the subsurface, it is a major aquifer used for drinking water in Northern Illinois. Water wells typically produce high volumes of clean water. Aquifers are geologic units or combinations of units, which permit the movement of groundwater. Groundwater almost never exists as underground rivers, especially in Northeastern Illinois. Instead, the water fills pore spaces (as in the case of sandstone) or cracks (as in the case of carbonate rocks). Shale tends to be impervious, and yields little to no water.

GPS Location: 41.3290 -88.9129 Photograph showing cliffs of Saint Peter Sandstone along the entrance road to Buffalo State Park, looking south.

GPS Location: 41.3258 -88.9096 Photograph showing the iron stained colors of the Saint Peter at the Illinois River in Buffalo State Park, looking west.

Illinois Route 71 and Illinois Route 178 (IL-71 & IL-178): Starved Rock State Park

One of the most visited state parks in Illinois is Starved Rock. The park has a rough history. The Native Americans populated the area almost continuously from 8,000 BCE until the middle 1800s. The actual rock itself gets its name from a battle between tribes in 1769. During the battle, the Illini Tribe was forced upon the rock by the Pottawatomie. Although the Illini brought enough provisions for several days, they spent more time on the rock than they had hoped. After provisions ran out, the story goes they starved to death rather than surrender. This is where the name Starved Rock comes from.

In the 1890's the land that would eventually become Starved Rock State Park was purchased by Daniel Hitt, who wanted the land for tourism. In the 1930's the trails that presently snake through the park were cleared. In 1966 the park became a National Historic Landmark. When you enter the park, the first manmade structure that you come upon is the visitor center. This is an excellent place to learn about the history, prehistory, flora, and fauna of the area.

GPS Location: 41.3207 -88.9931
Photograph showing the entrance to the Visitor Center, looking southeast.

Starved Rock State Park sits on the southern side of the Illinois River, less than a quarter of a mile west of the Lock and Dam. The park boasts excellent and scenic views with more than one hundred feet of relief. There are many small canyons carved into the rocks in the park. The vast majority of the park is made up of the Saint Peter Sandstone. The Starved Rock Member of the Saint Peter Sandstone gets its name from the park and is the most extensively exposed member in the park. The Starved Rock Member is white clean quartz sandstone deposited in a barrier island complex, similar to modern day coastal barrier islands in the Gulf of Mexico.

GPS Location: 41.3212 -88.9899
Photograph showing the Lock and Dam from the top of Starved Rock, looking east-northeast.

The Saint Peter forms such dramatic cliffs in the area for several reasons. In most places, the sandstone is not cemented well. It can be scratched with a dull knife. This makes it susceptible to surface erosion. Cracks in the rock will gradually widen until rounded high mounds are left. Glacial melt water also played an important role in the formation of features at Starved Rock. The Saint Peter is easier for water to cut than harder carbonate rock. This type of erosion forms majestic features in the park such as the falls at French Canyon.

GPS Location: 41.3177 -88.9912
French Canyon, looking east.

Illinois Route 178 and Illinois Route 351 (IL-178 & IL-351): Matthiessen State Park

Just west of Starved Rock lies Matthiessen State Park. Matthiessen is not visited as much, but from a geological standpoint, it is far more interesting. The Saint Peter is present, as it is at Starved Rock, and forms the area known as the Dells. The overlying Ordovician Galena-Platteville is exposed along with Pennsylvanian shale, sandstone, and coal.

There is more going on here than just stratigraphy. Matthiessen sits on the edge of a structure known as the LaSalle Anticlinorium (or Anticline). Locally it is known as the Peru Monocline between the Illinois River and the town of Lowell. The LaSalle Anticline is one of the longest trending structures in Illinois. It stretches from Lee County and is continuous in a south-southeast direction all the way to the Indiana border in Crawford County. Matthiessen is one of the few places it is visible at the surface. Here the anticline is visible dipping 20° to 35° west along the Vermilion River. Here the Platteville Group plunges into the river. The rocks did not fall into the observed position. They were originally flat and later folded to their present position during the initial formation of the anticline in the middle Ordovician. As you head east away from the Vermilion River, you will notice that the rocks become more horizontal. The horizontal rocks to the east were not folded, and remain in their original position.

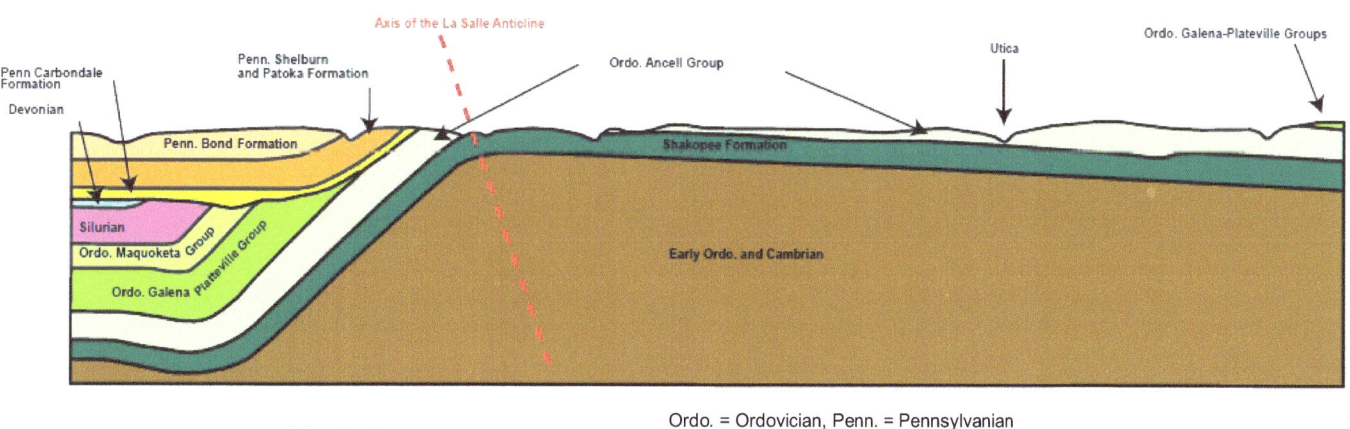

Ordo. = Ordovician, Penn. = Pennsylvanian

LaSalle Anticlinorium : *The LaSalle Anticlinorium is the major geologic feature in the area. This to scale cross section is along the Illinois River. The LaSalle Anticlinorium is locally called the Peru Monocline. The Anticlinorium began to form during the middle of the Ordovician. It later became reactivated during the Pennsylvanian. The location of many earthquakes in Northern Illinois occur on the anticline, indicating that it is active today.*

GPS Location: 41.2925 -89.0306
Photo shows the dipping beds of the Platteville Group, dipping west into the Vermilion River, looking north.

At the south end of Matthiessen you see the Vermilionville Sandstone at the Vermillion River. This sandstone is not as pure as the Saint Peter and contains much plant debris in the form of dark colored fossils and leaf imprints. It was deposited in ancient rivers during the Pennsylvanian. At a small creek that enters the Vermillion River, called Black Rock Run, you see other Pennsylvanian deposits. There is soft clay like shale and the black Colchester Coal is common. Coal is originally deposited as peat in coastal swamps. When people think ancient coal swamps, they usually think dinosaurs. These coals were deposited about 80 million years before the first dinosaur appeared. Black Rock Run is a narrow creek that squeezes through cliffs in the Galena-Platteville. Where it narrows, it is following fractures in the rock and occasionally carves out narrow passes.

GPS Location: 41.2810 -89.0294
The Vermilionville Sandstone in the top of a cliff face at a bend in the Vermilion River, looking northeast.

GPS Location: 41.2736 -89.0231
The Galena-Platteville dolostone at Reflection Pass in Black Rock Run, looking east.

There are also rounded almost black boulders, or nodules, of limestone. On a global scale, black limestone is rare. In the Pennsylvanian of Illinois, they are common. They are not glacially deposited. Glacial deposits in the park are restricted to the very top of hills. They were deposited in ancient Pennsylvanian seas by communities of microorganisms. These microscopic creatures incorporated some organic material giving the limestone its dark color.

GPS Location: 41.2736 -89.0260
Photo shows the black nodule boulders of limestone weathering out of the Pennsylvanian shale.

U.S. Route 6 & Illinois Route 178 (US-6 &IL-178): Abandon Canal Railroad Section

Railroad tracks offer some of the best outcrops within the State of Illinois. Most active tracks are dangerous and should be approached with caution. There is one place, just west of North Utica, where the tracks are no longer active. Exposed along the north end of the tracks, about three quarters of a mile due west of IL-178, is about 40 to 60 feet of shaley dolostone. This is some of the oldest rock accessible in Illinois. It is the Early Ordovician in age and is the Shakopee Formation. It is older than the Saint Peter Sandstone, which rests on top of it. An erosional unconformity separates the Saint Peter from the Shakopee. The outcrop can be viewed directly at the railroad tracks or from the Illinois Michigan Canal Trail south of the railroad tracks.

GPS Location: 41.3415 -89.0237
Photo shows the light colored dolostone of the Shakopee Formation, looking west.

The rock here is different from the later Ordovician and Silurian dolostone in that is contains small dome like structures. These small structures are composed of algal mounds called stromatolites. Stromatolites still exist today off the coast of Australia and are one of the oldest forms of life on Earth. They predate all large life by at least three billion years.

GPS Location: 41.3415 -89.0237 Photo shows a close-up of an algal mounds and mats (stromatolite) within the Shakopee Formation. The pencil is at the base of the mound, looking north.

Other small structures not common in dolomitic rocks are visible here. The rock contains a lot of shale. The shale particles help preserve things such as ripple marks, mud cracks, and low angle cross bedding. These features are usually only common in mudstones and sandstones. The fact that they are present along with the stromatolites tells us about the environment in which the Shakopee was deposited. Stromatolites live on tidal flats. Their presence along with the other small structures indicates that the Shakopee was deposited, at least locally, in a restricted marine lagoon that was highly susceptible to tidal influence.

Western Illinois

Galesburg Plain

The Galesburg Plain contains some of the most unique topography in the State. It was formed when the glacial ice of the Illinois Episode retreated over 100,000 years ago. The topography is older than what is observed in Northeastern Illinois. Here ridges are smaller, but tend to be more pronounced. They also tend to trend east-west as do rivers and streams. Streams contain deeper and wider valleys because they have had more time to erode the landscape. The surficial glacial deposits are mostly diamicton and loess. The exposed bedrock is significantly younger than in Northern Illinois. The major geologic structures present are the Mississippi River Arch, the Western Shelf, and the Glasford Impact Structure. Like the Des Plaines Disturbance (Volume I, p.49), the Glasford Structure is a meteor impact.

Griggsville Plain

The Griggsville Plain contains the oldest Illinois Age glacial deposits in Western Illinois. Here the glacial deposits are generally thin and overly unconsolidated bedrock that is either Carboniferous or Mesozoic in age. As a result, the topography is dominated by rolling hills. Here loess dominates over diamicton. The small Fishhook Anticline trends through the area.

Dissected Tills Plain

In Illinois, the Dissected Tills Plain is a thin sliver near the Mississippi River. It extends west into Missouri, where it consists of low cliff topography covered by rolling hills. In Illinois, Carboniferous bedrock and glacial loess forms most of the cliffs. The rolling hills are formed by the oldest glacial deposits in Illinois, dating back more than 750,000 years. The dominant surficial deposits are alluvial sediments from the Mississippi River. Streams within, tend to be small with deep valleys. The Mississippi Arch occupies most of the area.

Lincoln Hills

The southern part of the Lincoln Hills Section was never glaciated, just like the Wisconsin Driftless Section of Northwestern Illinois, and the Coastal Plain Province of Southern Illinois. The area is dominated by bedrock cliffs and flat plains. This variance in topography mainly formed as the modern Mississippi and Illinois Rivers do battle to occupy the area. Both rivers will naturally meander in their valleys. As they do, they laterally cut into hills and form high cliffs. The Capau Gre`s Faulted Flexure forms the southern part of this section.

Guide for Western Illinois

US-6 and I-88: Moline Area

As you enter Rock Island County from the east on I-88, you enter the Galesburg Plain. This area contains the transition from the Wisconsin Episode sand dune fields of Northwestern Illinois to flat outwash plains cutting across glacial diamicton of the Illinois Episode.

U.S. Route 6 and Interstate 88 (US-6 & I-88): Distal Ridge

As you head west on I-88 and approach the Iowa border, you will notice a conspicuous ridge on the north side in the distance. This ridge is most prominent from about mile marker 6 to mile marker 10. The ridge appears out of place, which has led many people to think of it as an esker. An esker is a linear ridge of sand and gravel on a flat plain. The ridge is made of finer grained sediments and isn't really a ridge. The north side, not visible from the highway, slopes much more gently. The ridge is a till and loess remnant left as floodwaters rushed west to the Mississippi River about 12,000 years ago. It formed in a similar manner as the Swallow Cliffs in Northeastern Illinois. Only here, the topography isn't as dramatic.

GPS Location: 41.5809 -90.2106 Photograph shows the apparent ridge compared with farm houses, looking northwest at mile marker 8 on westbound I-88.

U.S. Route 6 and Interstate 88 (US-6 & I-88): Mud Creek Outcrop

Off a little street (Co. Road 450E), about 370 feet south at the junction with US-6, is a small creek. If you park near the bridge and slowly descend east about 150 feet following Mud Creek, you'll come upon a small but significant exposure. The south wall of Mud Creek shows Pennsylvanian black shale (about two feet exposed) under grayish brown Illinois Age diamicton (the Hulick Member of the Glasford Formation). The two units are separated by a one foot layer of orange brown gravel and cobbles with some iron concretions. The gravel is at the unconformity between the Pennsylvanian and the Quaternary. The pebbles and cobbles are subangular and heavily stained. It more closely resembles preglacial gravels in Southern Illinois and may be stratigraphically equivalent to the Mounds and Grover Gravels.

GPS Location: 41.4539 -90.3477 Photograph shows the gravel layer between the Quaternary and Pennsylvanian. The orange backpack is on the gravel, looking southwest.

Not too far away on the north wall, the gravel is not present. Instead it has been eroded. The Hulick and Pennsylvanian deposits are separated by a soft black buried soil. The soil is at least one foot thick and maybe one of the oldest geosols in Illinois. Its actual age is not known, although it may be the Pike Geosol (Volume I, p.44).

GPS Location: 41.4539 -90.3477 Photograph shows the buried black soil under the lighter diamicton, with a quarter on the soil, looking northeast.

I-74/747: Peoria Loop

The loop of roads surrounding Peoria travel up and down cutting through glacial and bedrock exposures in the area of the Illinois River. The modern topography is similar to what it was before the glaciers covered the area. The most dramatic glacial materials in the area are made of windblown glacial loess, sometimes as much as 100 feet thick.

Interstate 74/474 Loop (I-74/474): South Airport Road Outcrop

This 25 foot outcrop is located along the Illinois 116 and South Airport Road Junction. In the summer it is almost completely concealed by brush. Here the outcrop is in the top half of the Pennsylvanian deposits. All the rock exposed here belongs to the Shelburn Formation. It is one of the few Pennsylvanian outcrops with abundant, but poorly preserved plant fossil debris in the sandstone. It is also a good exposure of concretions in dark shale. The sandstone is the Copperas Creek Member, and about six feet of it is well exposed at the top of the outcrop. The underlying, almost black shale is the Anna Member. It contains red hard iron concretions. It is only about five feet thick and contains white, chalky splotches. Below the Anna the outcrop is mostly covered, but poking through in a few spots is about 10 feet of an unnamed and unconsolidated, gray, sandy siltstone to shale.

GPS Location: 40.6826 -89.6565 Photograph shows the contact between the Copperas Creek Sandstone (light colored ledge forming rock) over Anna Shale (gray colored rock), with a pencil below an iron concretion, looking northwest.

Interstate 74/474 Loop (I-74/474): Bartonville US-24 Outcrop

On the west side of US-24 is about 20 feet of well cut, Pennsylvanian sandstone. The best place to park and observe the road cut is on the north end, on the southbound side. This sandstone is better indurated than most Pennsylvanian sandstones and has a smooth appearance. The stratigraphic position of the sandstone has not been worked out. There are no detailed well logs in the area to deduce an exact position. It is likely near the top of the Carbondale Formation or the base of the Shelburn Formation. If it is part of the Shelburn Formation, it is likely the Copperas Creek Sandstone. If you examine the sandstone closely near the base, you will see small, angular reddish orange crystals poking out of the rock. These are small crystals of iron minerals.

Also visible are very flat (or tabular) cross beds that dip north and thin beds of persistent dark colored siltstone within the sandstone. The siltstone dips about 21° almost due northwest. This sandstone and siltstone may actually represent windblown sands along an ancient beach as opposed to a marine or stream origin. The best lines of evidence for a wind influenced mode of deposition are the tabular cross beds and the lack of plant fossils.

GPS Location: 40.6220 -89.6558
Photograph shows the cross beds in the outcrop, looking west.

Photograph shows the flat nature of the sandstone under the siltstone and the small grains of an iron rich mineral, looking southwest.

IL-95 and US-24: Peoria to Mt. Sterling

Heading south from Peoria on US-24, you are following the eastern edge of Galesburg Plain. This 80 mile stretch is cut by narrow streams with relatively deep valleys more reminiscent of the topography of Southern Illinois. The glacial deposits here are older than in the northern and central part of the state. As a result, streams and rivers have had more time to erode the surrounding landscape. These deep valleys often expose the local bedrock. Most of the bedrock along this route is Pennsylvanian in age.

Illinois Route 95 (IL-95): Seville Railroad Section

There is a railroad exposure west of a little town called Seville off of IL-95. The outcrops along the railroad are one of the places that geology students are brought in order to expose them to the stratigraphy of Pennsylvanian age rocks. The Pennsylvanian is a unique system of rocks. It was deposited in environments from shallow marine to land deposits. The seas frequently rose and dropped by perhaps less than 100 feet in response to waxing and waning ice caps in the southern hemisphere. Illinois itself was situated near the equator. This led to the deposition of somewhat regular cycles of sediments (cyclothems) of sandstone, shale, limestone, and coal. The term cyclothem has slowly been falling from usage.

In order to visit the outcrop you can park off the north side of North Seville Road just north of the railroad tracks at GPS 40.4885° by -90.3514°, then head west on the north side of the railroad tracks. As you walk west for about 1,100 feet, you'll notice a black limestone in the railroad cut. This is the Seville Limestone of the Tradewater Formation. This limestone is similar to the nodules along Black Rock Run at Matthiessen State Park (p.45-47). The Seville pinches in and out within a mica rich fine sandy siltstone, which is extensively exposed along the railroad cut. These deposits are shallow marine at the leading face of a delta.

GPS Location: 40.4894 -90.3550 Photograph shows a close-up of the black limestone of the Seville along the Spoon River, looking north.

From here, if you turn southwest and cut through the trees, you come along the Spoon River. You will notice a pale colored rounded rock all around the north bank. This is the Bernadotte Sandstone of the Tradewater Formation. It was deposited in a river and beach system along an ancient seacoast. It is slightly older (perhaps less than 50,000 years) then the Seville Limestone. It was deposited at the base of a cyclothem. The Bernadotte is a moderately hard, pale yellow, fine grained, mica rich quartz sandstone the closely resembles the Vermilion Sandstone at Matthiessen State Park, except the Bernadotte is fossil poor. It tends to form cliffs along the Spoon River.

GPS Location: 40.4891 -90.3560 Photograph shows the Bernadotte Sandstone, looking east.

U.S. Route 24 (US-24): Duncan Mills Outcrop

This outcrop extends along the south bank of the Spoon River, right under the US-24 overpass. It is best observed from the top of the bridge, on the east side of US-24. Exposed here is the base of the Tradewater Formation. At the top of the outcrop is a thin layer, several feet thick, of pale brown glacial diamicton. The Bernadotte Sandstone (the same sandstone at the Seville Railroad Section) forms the top ledge, just below the diamicton, and is about two and a half feet thick. This outcrop is essentially what is buried beneath the Bernadotte Sandstone along the Seville Railroad Section (previous page). If you set the Seville Railroad Section on top of the Duncan Mills Outcrop, you would essentially have a continuous stratigraphic column of the Tradewater Formation. Immediately under the sandstone lies about 1.5 feet of the Pope Creek Coal, one of the thickest coals in the area. Most of the rest of the outcrop is unnamed soft gray shale. About halfway down the outcrop, nodular limestone is present in a three foot interval. The limestone is not named and appears to be a local limestone, confined to the outcrop area.

GPS Location: 40.3408 -90.1892
Photograph shows the Tradewater Formation in the south bluff of the Spoon River. The dark colored, thin ledge halfway up the outcrop is an unnamed limestone. Photo is looking east.

U.S. Route 24 (US-24): Ripley Section

This outcrop occurs on both sides of US-24 and contains such a wealth of stratigraphic, sedimentary, and environmental geology. Days could be spent here studying this outcrop. However, we are going to confine our description to the east side exposure for two reasons. Unlike the west side, the east side of the outcrop has not been engineered since it was originally cut and there is abundant parking on the east side.

GPS Location: 40.0272 -90.6312
Photograph shows the Tradewater Formation in contact with the Salem Formation.
Photo is looking southeast.

If you visit this outcrop on a hot summer day, you will notice the faint rotten egg smell in the air. This is sulfur and it is the bright yellow color in the outcrop. It is confined mostly to the shaley coal bed and sandstone at the top of the exposure. Along the south end, you will notice a pale yellowish gray rock that is highly stained red. This is the Mississippian Salem Formation, which is mostly a soft dolostone interbedded with gray shale. As you turn to the left, you'll notice that it disappears. What happened to it? Could there be a fault here? No. There are no faults in the area. The Salem Formation is missing to the north because after the Mississippian was deposited there was about ten million years of erosion that occurred in Western Illinois. The erosion happened when the seas withdrew and the entire Chesterian Series was eroded. During the Pennsylvanian, a stream flowing to the southeast carved a shallow valley into the Salem Formation. It was about the size of the modern La Moine River that is located less than 500 feet south of the outcrop. As the Pennsylvanian seas approached the outcrop from the southeast, the stream cut channel filled with a thin layer of gravel, sandstone, and coal. The seas deepened further eventually covering the outcrop and gray shale was deposited in a delta, followed by stream deposited sandstone at the top that fed the delta.

If you look closely where the Mississippian Salem Formation meets the Pennsylvanian coal, you will notice that the coal, about one foot thick, and drapes over the highest point of the Salem Formation. This drape deposit formed as the coal and shale were compressed with the weight of the above younger sediments. Shale tends to compress when a load is added, sand and hard rock tend to keep their original shape. The coal was originally deposited as peat when small plants died. We know the plants were small since the Salem Formation has no evidence of root activity. The coal contains a lot of clay. It has been described in the past as black shale, due to the high clay content. Also the underclay beneath the coal is very thin, lacks roots, and contains clasts of the Salem, indicating that soil development was poor when the peat was deposited.

The Pennsylvanian at the outcrop is the Tradewater Formation. The sandstone at the top is the Babylon Sandstone. It contains black plant debris, similar to the Copperas Creek Sandstone in Peoria. The deposits under the Babylon are local and have not been named.

GPS Location: 40.0272 -90.6312
Photograph shows the sulfur (yellow color) in the coal unit present in the channel fill, just to the upper left of the orange backpack. Photo is looking southeast.

GPS Location: 40.0272 -90.6312
Photograph shows how the coal drapes over the angular unconformity with the Salem Formation. The solid yellow line separates the Pennsylvanian (above) from the Mississippian (below). The dashed yellow lines show how the shaley coal "drapes" over the Mississippian. Photo is looking southeast.

IL-104 and I-72: Liberty to Griggsville

Liberty sits on the Dissected Till Plain. IL-104 heads east from here crossing the Griggsville and Galesburg Plain turning south along IL-107 for about 31 miles before reaching Griggsville and I-72. This area exposes thin glacial deposits over bedrock. The bedrock here is slightly different than in the surrounding areas. Mississippian and Pennsylvanian deposits are common but the area has other unique deposits. The Griggsville Plain is a high area where Cretaceous deposits are preserved. It is the only area in Illinois, other than the southern tip, where dinosaur age deposits are found. As you head south from IL-104 and eventually merge eastbound with I-72, glacial deposits once again dominate over bedrock exposures on the Griggsville Plain.

Illinois Route 104 (IL-104): Siloam Springs State Park, Campground Creek Section

Siloam Springs State Park is a semi-isolated park in Adams County. The nearest large road is US-24 about nine miles to the north. Isolated does not mean there is nothing to see. The park sits on Pennsylvanian deposits covered with thin glacial deposits. The actual geology is mostly covered by vegetation. Yet, there are many land features in the park. The park is host to Crabapple Lake. The lake is not natural. The creek was dammed in 1955 to form a 63 acre lake in the center of the park. The Campground Creek Section is one of the few exposures in the park. It is in the bed of a small unnamed tributary, east of the northern tip of Crabapple Lake. The exposed features only run along the small unnamed creek for about 500 feet. Near the south end of the section, weathered Pennsylvanian sandstones are visible poking through glacial sands. These sandstones are so weathered that they have almost become sand once again. You can tell the Pennsylvanian sands from the glacial ones mainly by the small iron concretions and finer grain size within the Pennsylvanian.

GPS Location: 39.8871 -90.9348
Photograph shows the weathered Pennsylvanian sandstone, looking west. A pebble sized iron concretion is visible at the top center left.

Visible near the north end of the exposure, the small creek is attempting to cut off one of its own meanders. It is a miniature version of what happens on a larger scale. As streams and rivers age they stop down cutting and start eroding laterally. As they do they develop meanders or bends. Sometimes a bend becomes so tight, that the river actually abandons a meander for a straighter path. Sometimes a cutoff meander becomes an oxbow lake. This stream is too small for an oxbow lake to form, but not too small to form a cutoff.

GPS Location: 40.8876 -90.9343 Photograph shows the active meander cutoff forming in the small unnamed creek, looking west. The present course of the creek goes to the left of the photo and comes back around the small sand hill in the center. Eventually the small sand hill will be eroded allowing the creek to flow in a straight line from the lower left to the upper right of the photo.

Illinois Route 104 (IL-104): Fishhook Creek/3000N Section

About three and a half miles south of Siloam State Park is a small creek. This is Fishhook Creek. You can park along the northeast side of the bridge over the creek, along E3000th Street at GPS: 39.8367° by -90.9405°. Here you can find large geodes in the creek bed. The geodes are within the Mississippian age Warsaw Formation.

GPS Location: 39.8368 -90.9410 Photograph shows large rounded geodes in the creek bed, looking west.

The Warsaw is dominantly calcareous shale with limestone. It is also very fossil rich. It contains rare fossils called *Archimedes*, which are bryozoans. They look like screws but are actually animals. They appeared around the time the Warsaw was deposited in a shallow sea, but went extinct around the time dinosaurs become abundant.

Photograph shows Archimedes fossil in the Warsaw. The Archimedes fossil is the large one at the center that looks like a line of small triangles in this specimen. Close to actual size.

On the west side of the bridge, there is a two foot layer of orange gravel, similar to what is at the Mud Creek Stop. These gravels are pre-glacial in age. They are likely Paleogene, but their exact age isn't known.

GPS Location: 39.8378 -90.9423 Photograph shows the Paleogene gravels (possibly the Mounds Gravel) at the top of the orange backpack, looking north.

Illinois Route 104 (IL-104): 2700E Section

On 2700E, about two thirds of a mile south of IL-104, on the west side of the road, is a small manmade gully. This gully is not impressive by any standard. It contains one of the best accessible exposures of Cretaceous aged deposits in Western Illinois. The Cretaceous is the only period in Illinois that represents the Mesozoic Era. The Mesozoic is also known as the age of the dinosaurs. No fossils of dinosaurs have ever been found in Illinois. The Cretaceous age deposits in this small gully belong to the Baylis Formation. It is mostly terrestrial stream deposits that formed in lowlands. When it was deposited, Illinois was high and dry. Although the nearby state of Iowa was occasionally covered in a shallow sea called the "Greenhorn Seaway", it doesn't appear to have encroached into Illinois (Volume I, p.35).

In Illinois the Baylis Formation is a silty to sandy clay that is almost white to dark yellow in color. It often contains orange iron staining. Its color distinguishes it from overlying glacial deposits, which tend to be greenish brown or gray in color. It forms the slopes of gentle rolling hills in the area. It is also thin, generally less than 30 feet and sits on top of harder, laminated Pennsylvanian shale.

*GPS Location: 39.8313 -90.9930
Photograph shows the white clay common in the Baylis Formation, looking southeast.*

Interstate 72 (I-72): Pike-Scott County Line Outcrop

On the north side of I-72 westbound, just west of the Illinois River, is a 700 foot long, 25 foot high outcrop. It was cut during I-72 construction, so it is relatively fresh. Even the two inch diameter core holes drilled to break the rock apart are still visible. This exposure is the Burlington and Keokuk Formations (often just referred to as the Burlington-Keokuk Limestone). The exposure here is well bedded, cherty, coarse grained light gray limestone interbedded with yellowish silty limestone. In many ways it resembles the older Silurian rocks of Northeastern Illinois. The Burlington-Keokuk Limestone forms prominent cliffs along the Illinois River Valley. It is Early Mississippian in age. Outside of North America, the Pennsylvanian and Mississippian are so similar that they are lumped together as one geologic period called the Carboniferous.

GPS Location: 39.6872 -90.6464 Photograph shows the light brown, coarse grained limestone in the Burlington-Keokuk, looking north-northwest. The white rock below the penny is a thick nodule of chert. Also visible is the remnant of a cylindrical drill hole made during I-72 construction, to the right of the penny.

Central Illinois

Ancient Illinois Floodplain

The Ancient Illinois Floodplain is sandwiched in a narrow strip between the Galesburg Plain to the north and the Springfield Plain to the south, and it follows the Illinois River valley. Until the end of the last ice age, the Illinois River did not exist. Instead, the Ancient Mississippi River passed through this area. The glaciers from the last ice age clogged the Ancient Mississippi River, forcing it into its modern valley. When the floodwaters of the Kankakee Torrent carved the Illinois River, the ancient course of the Mississippi River was exhumed becoming the southern part of the Illinois River. This is why the valley of the Illinois River is very broad and the landscape contains older geologic features. The dominant surface deposits are river alluvium with loess and some diamicton. The Sangamon Arch and Western Shelf are the dominant geologic structures in the area.

Springfield Plain

The Springfield Plain is similar to the Galesburg Plain of Western Illinois. Its topography is influenced by the deposits left by the ice sheets during the Illinois Episode. Flat, gentle rolling topography is the most common feature seen. The landscape is dotted with narrow high ridges that generally trend northeast-southwest. Some of these ridges are end moraines, others are eskers. Major tributaries often expose the bedrock within the plain. It is also some of the most fertile farmland in the world. Diamicton dominates the surface geology. The major geologic structure is the Sangamon Arch. The Fairfield Basin and Sparta Shelf also cross the area. The LaSalle Anticlinorium passes through the east end of the plain.

Kankakee Plain

Perhaps one of the most interesting geomorphic features in Illinois is the Kankakee Plain. The area it occupies was extensively covered by glacial ice. Yet, the glacial deposits are thin throughout. They are generally less than 50 feet thick compared to the surrounding areas where they are as much as 300 feet thick! The Kankakee Plain was heavily carved by glacial melt water after the glaciers receded. The melt waters that catastrophically carved the plain were part of the Kankakee Torrent that supplied massive amounts of water to the Illinois River from about 12,000 to 18,000 years ago. The height of the floodwaters occurred from 14,000 to 17,000 years ago. The result of these catastrophic flood events was the removal of glacial diamicton and the deposition of sand and gravel. Most of the hills throughout the plain are actually sand dunes that formed on the barren landscape. Outwash sands are common along with loess, sand dunes, and thin diamicton. The Kankakee Arch passes through the area into Indiana. The Sandwich Fault Zone also terminates within the Plain in Will County.

Bloomington Ridged Plains

The Bloomington Ridged Plain shows features somewhat in between what is observed in the Wheaton Morainal Country and the Springfield Plain. The surface glacial deposits were deposited by the Wisconsin age glaciers, which receded from the area about 16,000 years ago. They left behind a young landscape of rolling hills and shallow rivers. As the glaciers retreated north, they left behind several short lived glacial lakes in the counties bordering the Kankakee Plain. These lakes all drained during the late stages of the Kankakee Torrent. Due to the youthful topography and thick glacial deposits, streams and rivers have not had enough time to crave deep valleys. As a result, there is very little bedrock geology exposed at the surface. The LaSalle Anticlinorium is the dominant bedrock structure to occur in the area.

Guide for Central Illinois

IL-18 and IL-26: Henry Area East of the Illinois River

The area of Henry Illinois is located on the Bloomington Ridged Plain. The area is cut by the Illinois River which provides some unusually good exposures of glacial features not seen in the surrounding plains. The area is full of small streams that meander and jog their way through to the floodplain of the Illinois River. The floodplain is far wider than anything the modern Illinois River could have cut. This north-south stretch of the Illinois River, beginning just 13 miles to the north, is where the Ancient Mississippi River once flowed until its course was altered by the Wisconsin Episode glaciers.

Illinois Route 18 and Illinois Route 26 (IL-18 & IL-26): Sandy Creek at IL-18 Bridge

Sandy Creek occupies the floodplain of the Illinois River. At one time the Illinois River occupied the area during the Kankakee Torrent. Sandy Creek is by no means a mighty river, but it has influenced the local surface geology. One recent example of this influence can be seen under the IL-18 Bridge. Here the floods of 2011 have deposited and eight foot high debris pile at the upstream side of the bridge. You can park on the southwest corner of the IL-26 and IL-18 junction (GPS: 41.0987° -89.3403°), then head west towards the bridge. The debris was still present in the summer of 2012 when I visited the location. The result of this unintentional dam is causing Sandy Creek to meander around the bridge to the west and the east, widening the stream. This can potentially weaken the footings if it is not removed. These natural dams caused by manmade structures are common during flood events.

GPS Location: 41.0990 -89.3422
Photograph shows the debris dam at the IL-18 Bridge, looking north.

Illinois Route 18 and Illinois Route 26 (IL-18 & IL-26): Sandy Creek North Section

About two thirds of a mile east-northeast of the previous location, on the north side of road 1245N is a cliff composed of unusual material. Most cliff faces in Illinois are usually held up by bedrock or glacial loess deposits. This top half of this outcrop is composed of glacial outwash deposits of the Pearl Formation. The Pearl Formation is from the Illinois Episode of glaciation and varies in age from about 125,000 to 185,000 years ago. At this location is comprised mostly of well bedded to cross bedded conglomerate that has been somewhat consolidated. The result is the 80 foot tall cliff exposed at this location. The conglomerate was deposited by fast moving water in glacial outwash streams. This exposure was not quarried by man. It is a natural cut eroded by the Ancient Mississippi River when it occupied this part of the Illinois River Valley now occupied by Sandy Creek.

GPS Location: 41.1006 -89.3301 Photograph shows the bedding visible in the Pearl Formation near the top of the outcrop, looking west.

GPS Location: Photograph shows the variety of rounded clasts in the conglomerate of the Pearl Formation, looking northeast.

I-55: Bloomington to Springfield

The journey south from Bloomington to Springfield is generally one of low rolling hills and flat farmland with small areas of end moraines that are quickly passed by. Beneath these features lies mostly glacial diamicton from the Wisconsin and Illinois Episode glaciers. Although the area is extensive, it hasn't been studied in much detail due to poor surface exposure. Yet the area is extremely diverse in glacial deposits and land features. Glacial lake beds are common as are buried rivers, end moraines, and poorly understood ridges. Glacial deposits are also at their thickest along the highway, reaching as much as 500 feet in thickness.

Interstate 55 (I-55): Funks Grove Rest Area, Southbound

As you travel south on I-55 from Bloomington for 60 miles to Springfield, you are driving on top of the oldest Wisconsin age diamicton, the Tiskilwa Formation. There aren't many good exposures of geologic features. Fortunately, human beings often move things to places where you can see some geology. One such place is the southbound Funks Grove Rest Area. This rest area has a lot of local displayed history inside the main building.

Outside the building, there are light gray boulders that have been placed along the side of the road. These boulders are limestone and were likely quarried from local Silurian reef rock. If you look closely at them you will see that they are loaded with fossils and white chert. The fossils are mostly worm burrows with algal *Receptaculites* fossils.

GPS Location: 40.3576 -89.1105
Photograph shows the fossils within the boulders at the side of the rest area ramp. The fossil to the right of the quarter is likely Receptaculites.

Interstate 55 (I-55): Elkhart Hill

Off of the east side of I-55, in Logan County, sits a precarious hill just to the east, outside of the town of Elkhart. This hill is Elkhart Hill. It stands about 200 feet higher than the surrounding area and parallels I-55 for about one and a half miles. This hill has traditionally been thought of as an esker left by the Illinois glaciers. Recently this interpretation has come into question. There are no detailed well logs on the hill itself. There are small outcrops on the top that look like diamicton, not sand and gravel, which would be expected if it were an esker. There are surrounding wells off of the hill which penetrate about 15 to 60 feet of the Radnor Member of the Glasford formation. Underneath the Radnor is about zero to 20 feet of what appears to be the Teneriffe Silt which may be lake deposits. You don't start to see sand and gravel until you penetrate the Teneriffe, where at least 90 feet of the Pearl Formation is present. If the hill itself is diamicton, than it is possible that it formed where the sides of two separate glaciers met. This is called an interlobate area. Until there are detailed deep wells or excavations conducted on top of the hill, its exact origin will remain a mystery.

GPS Location: 40.0201 -89.4950
Photograph shows Elkhart Hill, looking east from Kennedy Road.

I-57: Kankakee to Champaign

The city of Kankakee sits firmly within the Kankakee Plain, hence the name. As you begin your journey south on I-57 for about 75 miles, you will notice that the area has small hills surrounded by flat land. These small hills are mostly sand dunes on top of glacial outwash, left by the Kankakee Torrent. The glacial deposits here are a thin coat on top of the Silurian carbonate bedrock. As you enter Iroquois County from Kankakee, the landscape is similar to the stretch along I-55.

Interstate 57 (I-57): Mahomet Bedrock Valley

Along I-57, near Paxton (just south of the Iroquois/Ford County border) is an area of farm fields. These flat and featureless fields cover something very significant. About 250 to 300 feet below the surface is a bedrock valley filled with sand and gravel. Although the valley is near 300 feet at its deepest, it is completely covered with glacial deposits. This is the Mahomet Bedrock Valley. It trends east-west before turning southwest. Before the glaciers initially advanced into Illinois over one million years ago, the Mahomet Valley contained a river that rivaled the Illinois and Mississippi rivers of today. The valley's has its beginnings in West Virginia and crosses the entire states of Ohio and Indiana before entering Illinois at the southeast corner of Iroquois County. It eventually links up with the modern Illinois River near Peoria. As you drive past Paxton Illinois try to picture driving across a river valley that was once 300 feet deep and about two miles wide. Although the valley is deeply buried today, it still contains a lot of water in the subsurface. Water wells set into the sands of the Mahomet valley are a major source of groundwater for the eastern part of Illinois.

Interstate 57 (I-57): Illinois State Geological Survey (ISGS) of the Prairie Research Institute

East of I-57 and south of I-74 is home to the University of Illinois in Champaign. Within this large campus is the home of the Illinois State Geological Survey. The main building is located at 615 East Peabody Drive. It is here that most of the professionals who decipher the geology of Illinois do their work. It is also a great place to see geologic samples from all over Illinois. Any of the staff would be happy to guide you around the building.

Not everything is housed at 615 East Peabody. There is a large warehouse of rock and core samples located in the Samples Library at 1910 Griffith Drive in Champaign, almost exactly one mile southwest of 615 East Peabody. The Samples Library is open to the public, and is at present free of charge. In 2008, the ISGS was incorporated into the Prairie Institute, along with four other surveys. This transferred it from the State of Illinois to the University of Illinois.

GPS Location: 40.1013 -88.2290
Photograph of the Illinois State Geological Survey building on 615 East Peabody, looking east-southeast.

Photograph is inside of 615 East Peabody. This is a piece of petrified wood collected from the Cretaceous sediments of Alexander County.

GPS Location: 40.0201 -89.4950
Photograph was taken outside of 615 East Peabody. This large glacial erratic was brought down from the Sault Ste. Marie area of Ontario, Canada by the most recent ice age. It is originally from the Precambrian aged Gowganda Formation. It was deposited as diamicton, like most glacial deposits in Illinois. The only difference is that this hard piece of glacial diamicton is 2.4 billion years older than the glacial deposits of Illinois.

GPS Location: 40.0913 -88.2426
Photograph is inside of the Samples Library, showing the warehouse full of boxed and cataloged samples.

GPS Location: Photograph was taken outside of 615 East Peabody during a temporary display. It shows a two inch diameter rock core drilled into the Pennsylvanian.

I-74: Bloomington to the Indiana Border

If you choose to travel east ion I-74 from Bloomington, the terrain appears different. The change in topography is a matter of perspective. Instead of crossing perpendicular to end moraines, like you do on I-55 and I-57, you are driving on them for the most part. This makes the topography appear more ridged. Bedrock exposures are rare but some do exist near the Indiana border.

Interstate 74 (I-74): Moraine View State Recreation Area

As you head from Bloomington-Normal east on I-74 towards Champaign-Urbana, you will notice that the flatness of the area is broken up by these long east-west trending hills. These are end moraines. They are areas where the glaciers (to the north) parked themselves for a time. Although the ice within the glacier was still moving forward, melting back of the glacier equaled the advancement of the ice. It was in a state of equilibrium, perhaps for several hundred years. This resulted in the deposition of the Bloomington End Moraine and associated smaller moraines of the early Wisconsin Episode.

Moraine View State Recreation Area is located on an end moraine. The lake at the center of Moraine View is Dawson Lake. Dawson Lake was created with the damming of the North Fork of Salt Creek. It opened to fishing in 1963. The recreation area itself loops around Dawson Lake. The main entrance is at the west side where the T-junction of County Road 900N and Moraine View Park Road (GPS: 40.4133° by -88.7268°).

If you ever get the opportunity to camp here take a moment to stop and look south at the flatter area. Picture a barren plain with little vegetation and howling winds. Then face north. Imagine that a 500 to 700 foot thick glacier stands in front of you! This was the scene 20,000 years ago. Less than 35 miles to the southwest was the maximum extent of the Wisconsin Episode glaciers. See Volume I (p.43) for a block diagram of the local geology.

Interstate 74 (I-74): Kickapoo State Park

Near the Indiana border off of I-74, in Vermilion County, is Kickapoo State Park. Kickapoo is presently known for its winter fishing. It was designated as natures preserve in 1974, and is the only known home to the silvery salamander in Illinois. At one time the Pennsylvanian deposits were mined here for their coal between 1850 and 1940. Lakes now fill the areas that were mined out. Small Pennsylvanian sandstone bluffs greet you at the entrance to the park. As you navigate through the park, you will see other exposures, mostly of Pennsylvanian shale.

*GPS Location: 40.1393 -87.7526
Photograph was taken near the entrance to the park. The sandstone is likely the Trivoli Member of the Patoka Formation and is Pennsylvanian in age. Photo is looking south-southeast.*

GPS Location: 40.1286 -87.7357
Photograph shows the gray shale of the Pennsylvanian under a thin tan layer of glacial deposits about 475 feet due north of the I-74 overpass, looking south-southwest.

IL-100: Northern East St. Louis Area

The area surrounding East St. Louis is a rare meeting of several geologic provinces. The northern end sits firmly on the Springfield Plain that continues to the Mississippi River. The Mississippi River exposes much of the bedrock under the glacial deposits.

Illinois 100 (IL-100): Alton Bluff Section

On the north side of the Mississippi River, just west of the town of Alton, are almost white colored cliffs on the north side of IL-100. The bottom half of the cliffs are the Mississippian age St. Louis Limestone. The top half belongs to the Ste. Genevieve Formation. The limestone commonly forms bluffs along the Mississippi and Ohio Rivers. It is very cherty and it contains dolostone and anhydrite within breccia. It also contains abundant coral fossils. This stretch of outcrop contains 18 recognized faults within a 900 foot stretch. The faults are small and have a net displacement of five to six feet. They are likely the northern tip of the Saint Louis Fault which trends south-southwest from this point into Missouri for about 25 to 30 miles. The railroad tracks at the base of the cliffs are active. It is best to view the outcrop from a distance.

GPS Location: 38.8914 -90.1919
Photograph shows the massive, light colored cliffs of the St. Louis Limestone (below the dashed yellow line) and Ste. Genevieve (above the dashed yellow line) along IL-100, looking west-northwest.

Southern Illinois

Mount Vernon Hill Country

The Mount Vernon Hill Country is the largest physiographic division in Southern Illinois. As you head from north to south, you will notice that the typical flat topography of central Illinois is gives way to rolling hills. Streams in this area tend to drain in a more southern direction than they do to the north. This is partially due to Illinois Episode ice. The Illinois Episode glaciers retreated in a more northern direction than the later Wisconsin Episode glaciers. The Wisconsin Episode glaciers never made it this far south. The forces of erosion have had more time to remove glacial deposits here than in the other parts of Illinois. Diamicton covers the area but it is thin. The major geologic structures are the Fairfield Basin (Volume I, p.13) and the Sparta Shelf. There are also several large fault zones in the area, such as the Cottage Grove Fault Zone which cuts across the area in a roughly east-west line. A buried crypto-volcanic dome called the Omaha Dome exists in the subsurface of Gallatin County.

Shawnee Hills

The Shawnee Hills Section is more reminiscent of topography in Kentucky than in Illinois. It consists of large rolling hills and deeper valleys than the northern two thirds of Illinois. The bedrock in this area tends to poke up through the glacial deposits of the lowlands. This is because the glacial deposits here are extremely thin and some of the oldest in the state. Most of the glacial deposits consist of lake sediments and loess with only a thin mantle of diamicton. The diamicton left in this area by the Illinois glaciers has been eroded significantly from when it was deposited over 150,000 years ago. As a result, the bedrock is extensively exposed throughout the area and karst topography is common.

Beneath the surface the Illinois Basin is at its deepest here. Here the Precambrian rocks are more than 14,000 below the surface; as compared to places in Northern Illinois where the Precambrian is about 1,500 feet below the surface. The Shawnee Hills Section is home to countless faults and small structures. The major fault zones are the Rough Creek Fault System and the Wabash Valley Fault System. Hicks Dome and the related Tolu Arch are also present in Hardin County. The area is covered with thousands of sinkholes throughout the area (Volume I, p.17).

Coastal Plains Province

The Coastal Plain is the very northern tip of a physiographic province that extends south to the Gulf of Mexico. The area was never covered with glaciers, although river deposits, outwash, windblown sediment, and lake deposits are common. What we now call the Gulf of Mexico covered this part of Illinois about 50 to 70 million years ago. It also contains the thickest deposits of Cretaceous and non-glacial Cenozoic sediments in Illinois. Some of the Cretaceous sands in the Coastal Plain are white sand beach deposits that strongly resemble other deposits in Georgia and Alabama. The Mississippi Embayment is the dominant geologic structure in the area, along with many faults.

Salem Plateau

Salem Plateau Section is a narrow area covering small western parts of only five counties. It consists mainly of Paleozoic bedrock hills surrounded by flat alluvial deposits of the Mississippi River. Like the driftless area in Wisconsin, the topography on the Salem Plateau is older than in surrounding areas. This small stretch of land has been shaped by the Mississippi River since before the Quaternary began 2.6 million years ago. The dominant structures in the area are the Waterloo-Dupo Anticline and the Ste. Genevieve Fault Zone.

Guide for Southern Illinois

I-57: Mount Vernon to Cairo

As you head south from central Illinois along I-57 from Mount Vernon to Cairo, you will notice a distinct transition in topography. As you head south, there is less and less glacial cover as you travel the 95 miles. The glacial deposits present are some of the oldest in the state. Mount Vernon sits atop Illinois Episode sediments. As you approach Cairo, the rocks generally become older until you reach the Coastal Plain. Cairo itself actually sits near the convergence of the mighty Mississippi and Ohio Rivers. This landscape has only been shaped by running water within the past 6,000 years or so. Before that time, the Ancestral Ohio River flowed from east to west, as it does today, but it was significantly north of Cairo.

At the north end of the outcrop at on the east side of I-57 (GPS: 37.55073° -89.03978°) are small growth faults in the area where the shale transitions to the Cedar Creek Sandstone. Although the growth faults are no more than a couple of feet high, this location is the only easily accessible outcrop to see them in the entire State of Illinois.

Interstate 57 (I-57): Johnson-Union County Line Section

At the border of Johnson and Union Counties, on I-57, is a 20 foot outcrop on both sides of the expressway. The outcrop is just south of thee Goreville exit near mile marker 39 at the Johnson/Union County line. I-57 slowly descends as you drive south along the highway. This 1,800 foot long outcrop takes you back through time. The rocks at the south end of the outcrop are slightly older than the rocks on the north side. They are all Pennsylvanian.

The tan colored sandstone rocks at the north end belong to Cedar Creek Lens of the Tradewater Formation. The long gray stretch in the middle is sandstone and shale of the Jims Hill Member of the Tradewater Formation. The tan colored sandstone on the south end of the outcrop belongs to the top member of the Caseyville Formation, called the Pounds Sandstone. This entire outcrop represents less than one million years of deposition and shows a terrestrial environment (south end) transitioning to a near shore delta environment (middle) before returning to a terrestrial environment (the north).

At the north end of the outcrop at on the east side of I-57 (GPS: 37.55073° -89.03978°) are small growth faults in the area where the shale transitions to the Cedar Creek Sandstone. Although the growth faults are no more than a couple of feet high, this location is the only easily accessible outcrop to see them in all of Illinois.

GPS Location: 37.5503 -89.0405. Photograph is a panoramic view of the west side of the outcrop, north is on the right.

Interstate 57 (I-57): Highway Line Section

Along the east side of northbound I-57, just north of mile marker 27, is an exposure of about 30 feet of the Fraileys Shale and Haney Limestone. The Fraileys and Haney are members of the Mississippian Golconda Formation. There are several distinct lithologies exposed. At road level, in the drainage trench and up for about 12 feet, you see micaceous greenish shale, with some laminated sandy oolitic limestone. This is part of the Fraileys Shale. Above the shale is about 10 feet of gray oolitic limestone with discontinuous dark gray shale interbedded. This middle part is the Haney Limestone and is very fossil rich, containing large scale cross beds. At the top of the middle unit, about a foot or so above a two foot thick greenish gray sandy siltstone bed, is a diastem unconformity. The rocks above here are lighter packstones and are more evenly bedded. They also weather a yellow brown color. The rocks above the diastem are also the Haney Limestone. This diastem probably represents a period of depositional hiatus not of erosion or soil development.

*GPS Location: 37.4115 -89.1545
Photograph shows the entire outcrop, looking northeast.*

*GPS Location: 37.4115 -89.1545
Photograph shows large scale cross bedding present in the bottom of the middle unit, looking east. Cross bedding in carbonate rock is relatively rare and indicates a high energy shallow marine environment.*

GPS Location: 37.4115 -89.1545 Photograph shows a close-up of the oolite beds in the top half of the middle unit, looking northeast.

GPS Location: 37.4115 -89.1545 Photograph shows the diastem unconformity between the middle and upper lithologies, looking northeast. The quarter is on the diastem.

US-51: Carbondale to Anna

US-51 trends north-south from Carbondale to Anna for 19 miles, before merging with I-57 just north of Dongola. This stretch of highway sits just to the east of the Salem Plateau. The result is high topography with rolling hills of farmland. The higher topography tends to be ridges and hills of Mississippian or Pennsylvanian bedrock. The rolling hills are mostly shale covered by thin loess deposits. The area is also heavily wooded since a large part of it is within the Shawnee National Forest.

U.S. Route 51 (US-51): Cobden School Road Outcrop

As you head south on US-51, you will come across a series of outcrops beginning about 2.3 miles south of the Jackson County and Union County Line. The first outcrop is a little more than 700 feet north of the US-51/Cobden School Road Junction. Here you will see the Mississippian Kinkaid Limestone. Here the Kinkaid is a sandy limestone separated in the middle by very calcareous shale.

GPS Location: 37.5663 -89.2324 Photograph shows the Kinkaid Limestone outcrop on the west side of US-51, looking northwest.

U.S. Route 51 (US-51): Blueberry Hill Road Outcrop

As you continue south along US-51 about 1.2 miles south of the pervious outcrop, you come up on the Kinkaid Limestone again, except it is cut by a different unit. The Pennsylvanian Keller Sandstone Lens (locally the base of the Caseyville Formation) sits on top of the Mississippian Kinkaid Limestone. Here the two are separated by an angular unconformity. Angular unconformities represent a period of erosion followed by a period of deposition. They are called angular because the beds below the unconformity dip at a different angle than the beds above. The difference in dips can be dozens of degrees or as little as a couple of degrees different. Here the Kinkaid dips less than 3° to the northeast. The Keller is flat. At the end of the Mississippian it was eroded to an uneven surface and later filled with sand that would later harden and become the Keller. The Keller Sandstone forms a strong contrast to the gray Kinkaid Limestone. The sandstone is clean coarse grained quartz sandstone. There is also a gray shale bed exposed at the base of the outcrop, within the Kinkaid.

GPS Location: 37.5503 -89.2254 Photograph shows the unconformity between the Kinkaid Limestone and the Keller Sandstone Lens, looking west. The Keller is above the yellow dashed line, the Kinkaid is below.

GPS Location: 37.5503 -89.2254 Photograph shows the coarse grains within the Keller Sandstone Lens, looking west.

GPS Location: 37.5503 -89.2254 Photograph shows the gray shale at the base of the outcrop. The limestone is at the top, the pencil is on the shale, looking west.

U.S. Route 51 (US-51): Bell Hill Road Outcrop

Further south along US-51, about one mile, just south of the junction with Bell Hill Road is an orange colored rock exposed on the west side. This is the Mississippian Degonia Formation, which is one formation below the Kinkaid Limestone. The Degonia contains sandstone in the upper and lower thirds. The middle is mostly limestone. Shale beds occur throughout.

GPS Location: 37.5354 -89.2215
Photograph shows the Degonia Formation on the west side of US-51, looking south-southwest.

I-24: Goreville to Metropolis

I-24 beaks off of I-57 about three miles north of Goreville. As you follow it south towards Metropolis for 31 miles (at the Ohio River), you pass through the high wooded hills of the Shawnee National Forest before you encounter the flatter Coastal Plain. Most of the bedrock exposed is Mississippian or Pennsylvanian in age. As you enter the Coastal Plain you get into beach sands and marine clays. Some of which are Cretaceous in age. The exposures of bedrock along I-24 are some of the most dramatic in Illinois.

Interstate 24 (I-24): Bowman Bottoms Section

About nine miles south of the I-24/I-57 split sits a 30 to 35 foot high outcrop on both sides of the expressway on I-24. This is one of the best exposures of the base of the Pennsylvanian Tradewater Formation anywhere in the State. The Grindstaff Sandstone forms the vertical cliffs. It contains purple iron strings that may someday form Liesegang Rings, if left to the forces of nature. The Gridstaff is chronostratigraphically equivalent to the Babylon Sandstone in Western Illinois, although the two do not physically connect. Tin this case it means they were deposited at the same time but not from the same source. Below the sandstone are piles of gray clay shale. These piles in front of the sandstone were not put here by people; they are natural outcrops that have eroded into mounds. Near the top is a black band less than one foot thick. It is the Reynoldsburg Coal. The clay itself has thin beds of fossil rich sandstone. Red iron concretions and plant fossils are common.

GPS Location: 37.4967 -88.8975
Photograph shows the massive brown Gridstaff Sandstone above the rounded clay and coal mounds, looking west. The arrows are pointing to the coal bed.

Interstate 24 (I-24): Odum Lane Section

About 10.5 miles south of the I-24/I-57 split, is an impressive outcrop that spans both sides of the highway. You would be very hard pressed to find another accessible manmade road cut of this magnitude in Illinois. It stands nearly 80 feet high on the northbound half of I-24. The sandstone is all the Battery Rock Sandstone of the Caseyville Formation. It is extremely hard and unique among Pennsylvanian Sandstones. Here it is conglomeratic. It contains rounded granular to fine pebbles of quartz, that become less abundant upwards. Their appearance is somewhat of an oddity as their source is hard to identify. It is possible that the pebbles were derived over a long distance from outside of the Illinois Basin. No one has done any detrital zircon analysis on the sand to determine a definite source area.

GPS Location: 37.4803 -88.8922
Photograph shows the orange weathering Battery Rock Sandstone, looking north.

GPS Location: 37.4803 -88.8922
A close-up of the quartz conglomerate at the base of the road cut, looking east.

Interstate 24 Route 146 (I-24 & IL-146): Vienna East Section

On the northbound entrance ramp to I-24 from IL-146, is an outcrop on the right side of the ramp. There is ample room to park. The outcrop is interbedded skeletal packstone (a limestone made of fossils) and gray to black shale. This is the Allard Member of the Mississippian Menard Formation. It contains abundant fossils in the shale. Well preserved *Archimedes* fossils over half an inch long are very common.

GPS Location: 37.4158 -88.8687
Photograph of the interbedding shale (recessed) and light colored packstone (protruding) of the Allard Member, looking northeast.

GPS Location: 37.4158 -88.8687
Archimedes fossils from the outcrop.

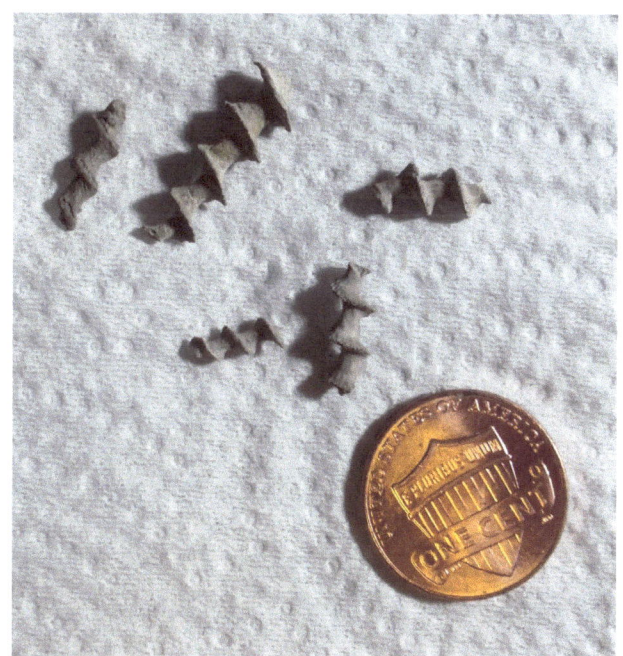

Interstate 24 and U.S. 45 (I-24 & US-45): Mallard Creek Section at Mount Pleasant Road

Eight miles northeast of Metropolis in Massac County, and six miles east-northeast of the I-24/US-45 junction lay a small creek that is usually dry during the summer. This is Mallard Creek. There is enough parking off the bridge that crosses the creek at Mount Pleasant Road, which crosses over the creek. About 200 feet west of the bridge the Cretaceous fine beach sands of the McNairy Formation are exposed in the walls of the creek. They are capped with Pleistocene Mounds Gravel. In some places the underlying black Post Creek Gravel (also called the Tuscaloosa Gravel), which looks almost like coarse asphalt, is exposed at the very bottom of the creek. To the east of the bridge, faulting is present that offsets the Cretaceous through Quaternary deposits.

GPS Location: 37.1504 -88.5848 Photograph of the fine sands and silts of the McNairy Formation topped with the Mounds Gravel, looking east.

Photograph shows the black gravel of the Post Creek Formation, looking south and down.

IL-1 and IL-146: The Ohio River to Vienna

Sitting at the very southeast corner, where IL-1 begins is a small town called Cave-In-Rock at the Ohio River you are presented spectacular views of the Ohio River from high bluffs. As you head north on IL-1 and then west on IL-146 you notice that there are many bedrock exposures and quarries. This area was extensively mined for fluorite, Illinois' state mineral. Abundant high angle faults cut this area and trend generally northeast from southwest. IL-146 roughly parallels the Ohio River from this point west until you reach Golconda, where it heads almost due west staying just north of the Coastal Plain. This 50 mile stretch is marked by subterranean caves and large sinkholes that have formed within the Mississippian limestone.

Illinois Route 1 (IL-1): Cave-In-Rock State Park

Further east along the Ohio River is Cave-in-Rock State Park. It is named for a large cave at the Ohio River. The cave is shrouded in tall tales of bandits, robbers, and lore dating back to the Revolutionary War. The cave has a very tall ceiling but narrow entrance. It is about 55 feet wide and has only the one entrance in and out. If you were a 19th century thief, it would have been a good place to hide out with a few horses for a couple of days, but is not the kind of place you want to get boxed into. The tales of the cave being used as a base of operations for bandits are likely exaggerated. The state acquired the land in 1929. The park has since grown to about three times its original size since the 64.5 acre acquisition in 1929.

Although it looks engineered, the cave was not carved out by man. It appears to have started out as a fracture in the St. Louis Limestone. Slowly rain and groundwater dissolved out the limestone around the fracture, slowly widening the cave, when it was strictly an underground feature. The present flat face of the cave was likely carved by flood waters of the Wisconsin Episode. Although the area was never covered by ice, the Ohio River served as an outwash channel for melting glaciers further north. If large rivers never cut out the cliffs; it is likely that the cave would have eventually developed into a sinkhole.

Photograph shows the inside of the Cave-In-Rock, near the back. Notice the light coming in from above at the top of the photo. This is probably where water began to percolate through the rock to slowly form the cave.

GPS Location: 37.4672 -88.1599
Photograph shows the entrance to Cave-In-Rock, looking north.

Illinois Route 146 (IL-146): Hastie Mine at Spar Mountain

Hastie Mine and Trucking is located just west of a large local structure called the Big Sink Sinkhole. As you enter the mine and head north you will drive over the St. Louis Formation and the Ste. Genevieve Formations. The north wall visible from IL-146 consists of all the members of the Paoli Formation and is capped by the Bethel Sandstone. All units are part of the Mississippian Mammoth Cave Group. This area has been locally mined for aggregate, lead, and fluorite.

GPS Location: 37.4971 -88.2037
Photograph shows the different Mississippian Formations exposed at Hastie Mine. The Paoli Formation makes up the visible gray rock in the wall. The brown rock at the top consists of the Downeys Bluff Limestone of the Paoli Formation which is capped by the Bethel Sandstone of the West Baden Formation. Photo is looking north.

Illinois Route 146 (IL-146): Town of Elizabethtown, Rose Hotel Outcrop

The Ohio River in Elizabethtown not only is the model for small town America, but it is also a place to view the erosive power of the water on bedrock. The outcrop is the Mississippian age St. Louis Limestone. The St. Louis is older than the Chesterian Series of the last stop. It was unquestionably deposited on the floor of a shallow sea. It often contains abundant brachiopods, crinoids, and corals. At this stop, it is accessible at the Ohio River, just south of the Rose Hotel Bed and Breakfast. This is private property, so you should ask before exploring the outcrop or approach it off property from the east.

Here we see the power that running water has on carbonate rock. The gray limestone is rounded as it slowly dissolves as rain and river water move across its surface. The pale yellow beds are chert. Chert does not react with the weak acids in water like the limestone does. As a result, it is far more resistant to erosion.

GPS Location: 37.4450 -88.3042
Photograph is the St. Louis Limestone overlooking the Ohio River, facing southwest.

Photograph is a close up of the gray rounded limestone interbedded with chert, looking north. The pencil is on top of the limestone.

Illinois Route 146 (IL-146): Pope-Hardin County Line Section

Along IL-146 about 1,100 feet west of the Pope/Hardin County Line is a 20 foot high, well faced road cut of what can be mistaken for limestone if viewed from distance. The outcrop's main face is actually calcareous sandstone. It is the Bethel Sandstone Member of the West Baden Formation. It weathers the light brown, but is grayer where not weathered. It is almost all fine grained. It also contains large brown lenticular nodules that are well cemented. There is gray sandy shale near the top above the recessed bench.

GPS Location: 37.4598 -88.4189
Photograph is a close up of the Bethel Sandstone, looking north. The pencil is on top of a brown concretion nodule.

IL-34: Rosiclare to Harrisburg

Rosiclare is a moderate sized town in Southern Illinois with a population of about 1,200 people on the Ohio River. As you head north, beginning your 30 mile journey along Main Street and link up with IL-34 (also shared with IL-146) the landscape is a mosaic of farms and forest, which cover the karst topography of the area. This scene continues as you head west on IL-34 just before it breaks off from IL-146 and heads north. As you continue north from the junction you will notice that the topography to the east significantly changes. There is a large hill that trends for miles. This is Hicks Dome. The dome is the dominant feature on the landscape. From this point north to Harrisburg, the landscape is cut by covered faults and karst land. Farmland gives way to the trees of the Shawnee National Forest.

Illinois Route 146 and Illinois Route 34 (IL-146 & IL-34): Hicks Dome

In Hardin County, nestled just northeast of the IL-146 and IL-34 junction, sits one of the most striking geologic structures in Illinois, Hicks Dome. The center, or apex, of the dome is near GPS 37.5313° by -88.3686°, about 0.9 miles south-southeast of Hicks just west of the Kaskaskia Experimental Forest. This feature is a 10 mile diameter, bull's-eye shaped structure, raised a couple of hundred feet above the surrounding landscape. The dome formed during the Permian Period when volcanic activity pushed up the rock from below. The magma that produced this feature, barely reached the surface in the form of rare thin dikes. The deep igneous activity displaced the local rocks by 4,000 feet upward! Devonian rocks are present at its center. The rocks become progressively younger away from the center.

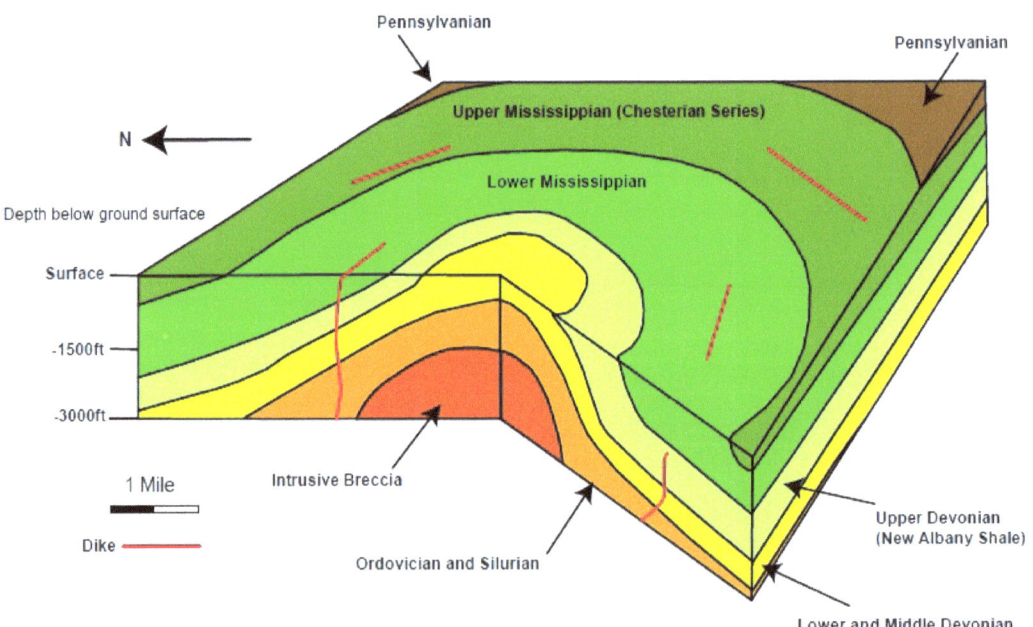

Hicks Dome: This simplified block diagram of Hicks Dome, shows how the Permian Intrusions deformed the rocks. Like all dome structures, the rocks become progressively younger away from the apex. Faults are excluded.

The dome is known to have been formed from deep igneous activity, yet the intrusion that formed it is somewhat of an enigma. The dome has been drilled into. The deepest well reached almost 3,000 feet terminating in rocks of the lower Ordovician but did not reach an igneous body. Usually, domes such as this have cooled magma under their center, as in the case of the Omaha Dome in the northwestern part of Gallatin County. The deep boring did yield volcanic breccia at 1,600 feet below the surface, some of which made it to the surface in volcanic vents. This indicates a very volatile magma that was rich in compressed gas. Faults form concentric rings around the dome and radiate from the center. The dome itself sits on the northwest end of the Tolu Arch which extends southeast into Kentucky. The two structures are physically connected and likely are related to one another.

Even with in this area of complex sedimentary and igneous structures, outcrops are rare. The dome is heavily forested and outcrops tend to show only a few feet of geology and most require long hikes to get to. It is visible from IL-34 as a large, broad hill to the east.

Illinois Route 34 (IL-34): Garden of the Gods, Shawnee National Forest

On the southeast corner of Saline County are strange rounded rocks in Garden of the Gods. Near Anvil Rock are large rounded rocks colored with circular features. These interesting features have formed on the rocks are the Pennsylvanian age Pounds Sandstone Member of the Caseyville Formation.

Sandstone often weathers into knobs but here something a little unusual is going on. The first things that you notice are the large circular shapes and rings on the surface of the rock. This is not bedding, they were caused by surface water eroding through cracks in the sandstone at a time when the sandstone was on flatter ground. These mini-structures are called Liesegang Rings. Groundwater percolating from the surface along with hydrothermal activity caused by Hicks dome just a couple of miles to the south, likely brought hot mineralized groundwater up from below when the area was tectonically active during the Permian. Unlike limestone which actually slowly dissolves in water, some of the sandstone gets carried off by the water. As a result the grains sometimes get stuck in the cracks. When this happens, water deposits iron around the edges of the weathered surfaces. The source of the iron was probably from hydrothermally enriched groundwater. As more time passes, the surrounding landscape erodes even further and some of the rounded boulders get eroded but the rings remain.

GPS Location: 37.6073 -88.3853
Photograph shows the rounded and hollowed out look of the Pounds Sandstone and Liesegang Rings, looking northeast. The arrows are pointing at the rings.

Photograph shows a close-up of the dark iron mineral patterns left by the iron rich waters that filled the cracks of the sandstone millions of years ago, looking southwest.

IL-3: Cairo to East St. Louis

IL-3 is one of the most geologically interesting highway in Illinois. As you head north from Cairo and follow it from the Coastal Plain into the Salem Plateau, you travel a landscape forged by rivers, oceans, and underground streams. This 130 mile journey boasts a landscape that has been 400 million years in the making. The east side is marked by high bedrock bluffs and the west side is occupied by the Mississippi River. The lowlands are often referred to as the "American Bottoms". From the ancient Devonian oceans to oxbow lakes less than 8,000 years old, the area holds a rich prehistory as well as paleo-history. The area's caves served as rock shelters for many native peoples as did the fertile lowlands. Even as you reach East Saint Louis, the high bluffs never surrender to the lowlands to the west.

Illinois Route 3 (IL-3): Horseshoe Lake State Conservation Area

Horseshoe Lake is a unique feature. Until about 6,000 years ago, the Mississippi River flowed through area. The Ancient Ohio River also flowed through the area up until about 8,000 years ago. The Mississippi River most likely diverted its course to its present location due to uneven subsidence in the area due to earthquakes. Horseshoe Lake remained an oxbow lake up until about 3,000 years ago when it became a swampy lake due to natural sediment filling. By about 1,000 years ago the lake was a swamp that was often dry in the summer. Native people used the area as fertile hunting grounds. In 1928, the State of Illinois created a dam on the south end of the oxbow lake. This allowed Horseshoe Lake to become a swampy lake once again. Alas, it won't last; eventually the lake will naturally fill with sediment and plant debris and will cease to exist.

At present, the lake supports a vast bio diverse flora and fauna. The most striking things that stand out are the bald cypress trees. The landscape is a land that time forgot. It more closely resembles the bayous and swamps of Louisiana then what we typically think of as Illinois. If this area were close to sea level, it might become buried by the meandering Mississippi River. After a few million years the peat would become coal. It would recreate an environment not seen in Illinois since the Pennsylvanian Period.

*GPS Location: 37.1226 -89.3340
Photograph is of the massive bald cypress trees that fill the southern half of Horseshoe Lake, looking east.*

Illinois Route 3 (IL-3): Rock Springs Hollow Section

About 1.6 miles west of IL-3 and 1.4 miles south of Thebes in Alexander County, sits a stream cut landscape taking you back to the Ordovician. It sits along Rock Springs Hollow Road, which is more easily accessed from the north-south trending gravelly Thebes/Fayville Road to the west. You can easily park on a small turn off 250 feet east of the bridge, on the south side of Rock Springs Hollow Road at GPS 37.1976° by -89.4488°. Just south of the parking area, is the Ordovician Girardeau Formation of the Maquoketa Group. It is a thin wavy bedded, fine grained limestone with chert nodules. The limestone weathers out easier than the chert. The chert will often form irregular blobs that stand out above the limestone. The vertical faces present in the creek look as if they were carved out to free the stone for building. They weren't. Natural joints and fractures (cracks), in the rock have gradually widened over time. Joints in limestone tend to form with two preferred orientations. One set of joints forms parallel to local structures, the other perpendicular to local structures. The gravel present in the hollow is redeposited Mounds Gravel from upstream.

If you follow the hollow downstream (under the bridge about 400 feet until you reach the abandoned railroad pillars) you will notice that the rock changes abruptly. You have just crossed the Rock Springs Fault. The fault trends north-northeast. On the east side you have the Ordovician Girardeau Formation, on the west side you have Silurian Sexton Creek Formation thrown down relative to the Ordovician. This whole area is dominated by small normal faults.

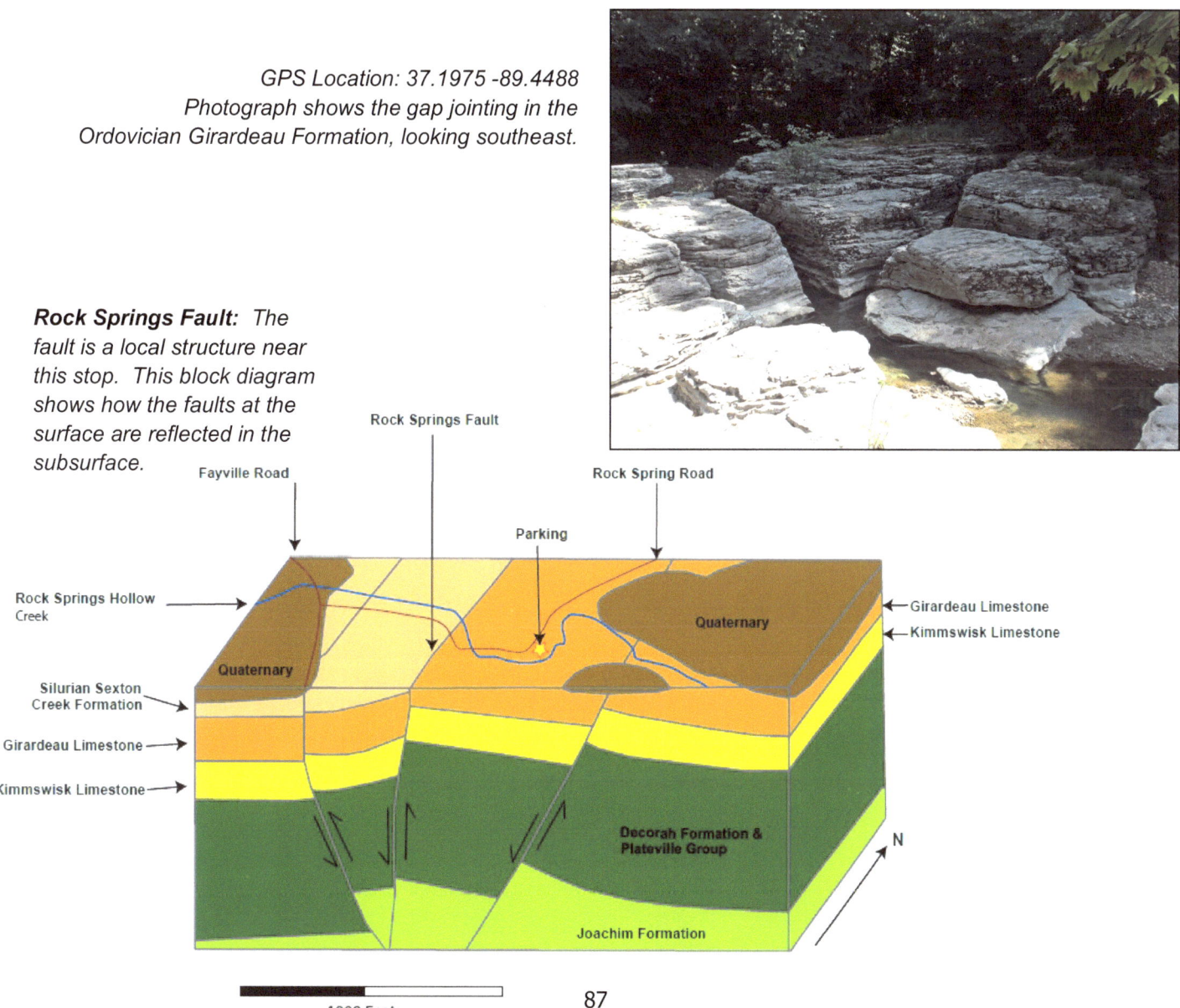

GPS Location: 37.1975 -89.4488
Photograph shows the gap jointing in the Ordovician Girardeau Formation, looking southeast.

Rock Springs Fault: *The fault is a local structure near this stop. This block diagram shows how the faults at the surface are reflected in the subsurface.*

Illinois Route 3 (IL-3): Thebes Rail Bridge Section

One and a half miles north-northwest of Rock Springs Hollow is a 230 foot high rail bridge crossing the Mississippi River on the south end of Thebes. This bridge was built in 1905 and is still used today. Just under the bridge and extending south for a few hundred feet is the Ordovician Kimmswick Limestone. This limestone is equivalent to all of the Galena Group in Northern Illinois, except for the Dubuque and Decorah Formations. The Kimmswick is a medium to coarse crystalline, high purity limestone. The upper part is argillaceous and contains brown to black chert nodules. Crinoids are abundant. This outcrop is best viewed in late summer to early fall, when the Mississippi River is low. You can park off the dirt road just north of the bridge at 37.2180° by -89.4623° and walk 400 feet south along the beach.

GPS Location: 37.2163 -89.4629
Photograph shows the low lying Kimmswick Limestone with Missouri in the background across the Mississippi River, looking south-southwest.

Illinois Route 3 (IL-3): Thebes Courthouse Outcrop

Along the bluffs on the east side of Thebes, just east of 4th Street, is a historical courthouse. This was the Alexander County Courthouse from 1848 to 1884, until it was relocated to Cairo. The building is now a historical landmark. It is one of the places where Abraham Lincoln practiced law.

The courthouse also sits upon the top of a geologically significant outcrop. This is the type section of the Thebes Sandstone, of the Ordovician, Maquoketa Group. The Thebes Sandstone is mostly a hard pale brown, fine grained, non-calcareous quartz wacke. It contains dark gray calcareous shale in the upper part, which is also exposed here. If you notice the dark shale dips sharply west near the top of the outcrop. The explanation for this sudden change in structure from flat to dipping 33°, is typical of drag folding along faults, but no fault has been documented here. The Thebes Sandstone is equivalent to the Scales Formation in Northern Illinois. The Thebes is the only persistent sandstone in the Maquoketa Group. It is found in the subsurface as far north as Brown County in Western Illinois. It was likely derived from the Ozark Mountain granites to the southwest in Missouri.

GPS Location: 37.2192 -89.4606
Photograph shows the Thebes Sandstone beneath the historical Alexander Courthouse. The lower light colored rock is sandstone. The dark gray rock above is calcareous shale that goes from flat to dipping 33° west, looking northeast.

Illinois Route 3 (IL-3): Pine Hills Escarpment

As you head north along IL-3 from Thebes, you start to notice bluffs to the east that slowly get closer to the road. About six miles south of Grand Tower just past Fountain Bluff, you will notice rock exposed in the hills. Here the bluffs are still more than a mile east of you but the rock exposure is noticeable, even in the summer. If you turn right (east) on La Rue Road (closed in the winter) and follow the gravel road to where it ends, you will be at the bluff exposure. You will have to get out and walk east to reach it. This is the Pine Hills Escarpment's southern edge. It trends north-south for 5,000 feet.

This cliff is a natural, 100 foot tall escarpment, originally carved by the Mississippi River and its tributaries. It is the best exposure of the lower Devonian Bailey Formation in Illinois. The bottom of the Bailey Formation was deposited during the Silurian. Deposition continued up through the beginning of the Lower Devonian. Unlike in Northern Illinois, deposition between the Silurian and Devonian was continuous in Southern Illinois. Here the Bailey is an argillaceous, very cherty limestone that exists in thin beds. The Bailey Limestone is one of the thickest formations exposed in Illinois at 500 feet.

GPS Location: 37.5456 -89.4391
Photograph shows the Bailey Limestone in the cliff face, looking north.

Photograph is a close-up of the chert in the argillaceous beds of the Bailey Limestone, looking east.

Illinois Route 3 (IL-3): Grand Tower Park, Tower Rock

Grand Tower Park is located about 1.5 miles west of IL-3 on the north end of the town of Grand Tower. The park is adjacent to the Mississippi River. In the Mississippi River stands Tower Rock. Tower Rock is a 90 foot high erosional remnant of the Lower Devonian Bailey Limestone. It sits in the middle of the Mississippi River and is technically on the Missouri side. It has served as a marker and risk to those navigating the Mississippi River.

GPS Location: 37.6352 -89.5109 Photograph was taken from a sandbar on the Illinois side of the Mississippi River looking southwest at Tower Rock.

Illinois Route 3 (IL-3): Grand Tower Park, Devils Backbone

The entrance road to Grand Tower Park cuts through a narrow north-south trending ridge. This Ridge is Devils Backbone and forms a feature referred to as a "hogback". A hogback is a ridge made up of steeply dipping beds of rock. The lower rocks belong to the Middle Devonian Grand Tower Limestone. The rocks at the top of the ridge belong to the Middle Devonian St. Laurent Formation. Devils Backbone is surrounded by subparallel faults to the west and east. There is an east-west trending fault exposed on the south side of Devils Backbone, outside of the park and another one on its northern tip.

GPS Location: 37.6390 -89.5108 Photograph is looking south at Devils Backbone. The rocks in the foreground (Grand Tower Limestone) dip 43° to the northeast, whereas the rocks at Tower Rock are flat. This indicates the presence of unexposed

Illinois Route 3 (IL-3): Fountain Bluff, Power Plant Road Outcrops

About 1.25 miles north of Grand Tower Park is the south end of a large elliptical bluff. This is Fountain Bluff and it trends roughly north-south for about four miles. On the south end, along Power Plant Road, sits several homes on top of the Mississippian Chersterian Series rocks. Behind the homes, in the high bluffs, are the lower sandstones of the Pennsylvanian Caseyville Formation. The sandstones are largely hard quartz arenites and are often seen being used as decorative stone in people's yards.

*GPS Location: 37.6528 -89.4998
Photograph is looking north at the sandstone bluffs along Power Plant Road.*

Illinois Route 3 (IL-3): Fountain Bluff, East Bluff Outcrops

As you circle Fountain Bluff and head north along IL-3, the Pennsylvanian sandstones are right at road level. If you look past the trees and well-manicured grass in the right of way, you will see the uneven bedding within the sandstone.

*GPS Location: 37.6811 -89.4782
Photograph shows the basal Pennsylvanian sandstones behind the trees along IL-3, looking west.*

Illinois Route 3 (IL-3): Rockwood Outcrops

The small town of Rockwood is located in the east bluffs along IL-3 in Randolph County. Here IL-3 is narrow, with few places to pull over. As you drive along it, you will notice outcrops on the east side. These outcrops are made of various formations within the Chesterian Series. On interesting rock is locally referred to as Swiss Cheese Block. It looks like two blocks of pitted cheese stacked on top of each other. The pitted appearance is caused by differential weathering. This occurs when flat faces of rock are eroded out creating large vugs. The rock is likely the top of the Palestine Formation. It can be hard to differentiate rocks in the area. They are poorly exposed and the Mississippian-Pennsylvanian unconformity runs throughout the area.

*GPS Location: 37.8382 -89.6961
Photograph shows Swiss Cheese Block, looking north-northeast.*

Illinois Route 3 (IL-3): Bluff Road, Mardoc Rock Shelter Natural Historical Site

About seven miles west of IL-3 and just north of the town of Mardoc, along Bluff Road, sits the Mardoc Rock Shelter archeological site on the north side of the road. The cliffs were used as camp sites by Native Americans from about 4,000 to 8,900 years ago. There are three geologic formations at this location. The basal part, that was used as a rock shelter is the Mississippian Aux Vases Sandstone. Higher up in the cliff is the Paoli Limestone. Up high in the bluff is the Bethel Sandstone Member of the West Baden Formation. All belong to the bottom of the Mississippian Chesterian Series.

*GPS Location: 38.0626 -90.0638
Photograph shows the sandstone and limestone at the rock shelter, looking east-northeast.*

Illinois Route 3 (IL-3): Bluff Road, Prairie du Rocher Outcrop

As you continue north along Bluff Road, you pass the junction with IL-155 in Prairie du Rocher. If you continue north on Bluff Road for about half a mile, you will see two large caves on the east side of the road. These caves are in the St. Louis Limestone. There is plenty of room to park along the east side. The caves are protected and entry is prohibited.

GPS Location: 38.09320 -90.1047
Photograph shows the caves within the St. Louis Limestone, looking east.

Illinois Route 156 (IL-156): Camp Vandeventer Section, Fountain Creek

This location is part of a Boy Scout camp. To get to it, you need to turn north onto Trout Camp Road from IL-156 and travel about one half mile north. The entrance sign to Camp Vandeventer will be on the east side of the road. Then you follow the winding road (generally north) until you reach the gate. Park, and walk north about another 800 feet until you reach Fountain Creek, the largest stream in Monroe County. You can't miss it. It is a 30 foot gorge cut into Mississippian bedrock.

The entire area surrounding the camp is known for its karst topography. This is one of the few places in Illinois where underground streams actually do exist. Streams feed the creek from upstream, pouring directly from caves in the bedrock. The rocks exposed are the Mississippian age St. Louis Limestone, the same rock unit exposed at Cave-in-Rock Park and Prairie du Rocher. Fountain Creek actually flows entirely underground during dry periods just a few hundred yards downstream of here. When there is a wet period, Fountain Creek is transformed into a raging river as it is fed by surface runoff and underground stream water.

GPS Location: 38.3421 -90.2037
Photograph shows the St. Louis Limestone exposed in the streambed and cliffs, looking east-southeast.

Illinois Route 3 and Illinois Route 158 (IL-3 & IL-158): Columbia Road Cut Section

As you head north on IL-3, approaching the IL-158 junction, you will notice a prominent 5 to 25 foot outcrop of rock on the east and west side of IL-3. The brownish colored rocks near the top are the bedded, cherty, and siliceous limestone of the Salem Formation. The gray colored rock beneath the Salem Formation is shaley limestone of the Warsaw Formation. Both units are lower Mississippian in age. The Warsaw Formation produces geodes in Fishhook Creek (p.58-59).

There is more than just Mississippian rock exposed here. As you pass the rocks, you are driving over a subtle, yet influential geologic structure. Here you are on the crest (or axis) of the Dupo-Waterloo Anticline. This anticline forms the border between the Salem Plateau Section (to the west) and the Mount Vernon Hill Country (to the east). As you head south on IL-3 and past the IL-158 junction, you will notice the beds appear to dip southeast. This is the flank of the anticline.

GPS Location: 38.4273 -90.1769 Photograph shows the beds of the Dupo-Waterloo Anticline about 360 feet north of the Hill Castle Road intersection with IL-3, looking south.

GPS Location: 38.4302 -90.1810 Photograph shows the Salem Formation on top of the Warsaw Formation at the entrance ramp onto IL-3 from IL-158, looking west-northwest.

Illinois Route 3 and U.S. Route 50 (IL-3 & US-50): Dupo to Cahokia Bluffs Outcrop

When you approach East St. Louis from the south, you will see high bluffs exposed from Dupo to Cahokia where IL-3 splits from US-50. Exposed in these bluffs is the St. Louis Limestone capped with about 20 to 30 feet of the Ste. Genevieve Formation. Caves and springs are present throughout the area. One excellent spring is located just northeast of Dupo and is called Falling Springs. Falling Springs flows all year round. It was originally used for water along the railroad. Today it is used to feed fish ponds on private property. Not only is there modern karst topography in the area, there are also small breccia filled paleo-karsts on top of the St. Louis Limestone.

GPS Location: 38.5323 -90.1917
Photograph shows the bluffs near Dupo, looking southeast. The photo was taken from Industrial Drive, a frontage road on the east side of the US-50 and IL-3 interchange.

Glossary

algae/algal – An organism or organisms comprised of marine dwelling single celled to complex, rootless, photosynthetic life/or having properties of algae.

alluvium – A deposit of sediments left by streams and rivers.

anhydrite – An evaporate mineral that occurs in layered deposits where large volumes of seawater have been evaporated, usually in a body of saltwater with no outlet to the sea.

anticline – A fold with layers of rock or soil sloping downward on both sides from a common ridge. Older rocks are at the center.

arenite – A sandstone or sand that contains particles of grains between 0.625 mm and 2 mm and contains less than 15% finer particles.

basement – Used to describe rock units that are below a sedimentary platform or cover and are normally ether igneous or metamorphic.

basin – A large depression that serves as the site for the accumulation of a large thickness of sediments. Beds dip towards the center from all directions. Younger rocks are towards the center.

bed/bedding – The characteristic structure of sedimentary rocks in which layers of different rock units are stacked one on top of another, in a layered sequence with oldest at the bottom and youngest at the top (in undeformed rock units).

bedrock – Rock units present beneath any surface soil, sediment or other surface cover. In some locations it may be exposed at earth's surface. In Illinois all pre-Quaternary geologic units are considered bedrock.

bituminous coal – A weak coal that contains a tar like substance. In hardness and quality, it is mid-grade.

breccia – Rock that is composed of large (over two millimeter diameter) angular fragments. The spaces between the large fragments can be filled with a matrix of smaller particles or mineral cement, binding the rock together.

calcite – An extremely common carbonate mineral and is a principle constituent of limestone and marble.

Cambrian – The first geological period/system of the Paleozoic Era of the Phanerozoic Eon.

carbonate rock – The group of rocks that contain limestone and dolostone.

Carboniferous – The fifth geological period/system of the Paleozoic Era of the Phanerozoic Eon. In North America it is split into the Mississippian and Pennsylvanian Periods/Systems.

cave/cavern – An underground void, formed naturally by a combination of processes ranging from chemical, erosion, tectonic, microorganism, pressure and atmospheric influences.

chert - A microcrystalline or cryptocrystalline sedimentary rock material composed of SiO_2, occurring as nodules, beds, and concretionary masses.

clay – A family of microscopic minerals formed from the alteration of other minerals, individual grains are less than 0.004 millimeters in diameter.

coarse grained – A descriptive word used to define rocks with particles sizes large enough to see with the naked eye.

coarse grained sand – Particles of rock between 0.5 and 1.0 millimeters in diameter.

coastal plain – An area of gently dipping rock units that were deposited near the ocean margin.

conglomerate – A sedimentary rock comprised mostly of rounded particles larger than 2 millimeters in diameter.

continental shelf – The gentle sloping margin of a continent that lies under the ocean.

convergent boundary – Plate boundary where old crust is being recycled into the mantle. They are often reflected on the surface as ocean trenches.

core – Located at the Earth's center and is divided into an inner core (solid metal), and an outer core (liquid metal). Both are composed mostly of iron and nickel. The Earth's core is about 16% of its total mass.

craton – A geologically inactive and relatively stable part of the continental crust.

Cretaceous – The third geological period/system of the Mesozoic Era of the Phanerozoic Eon.

cross beds – Beds of rock that were deposited at an angle other than horizontal. They can be flat or bowed.

crust – The solid and ridged outer shell of the Earth upon which all life lives. The two main types are the ocean crust and the continental crust. The crust accounts for slightly less than 1% of the Earth's total mass.

crypto-volcanic – A geologic structure, usually a dome, formed by subsurface igneous activity that only limitedly reaches the surface.

crystalline – A term used to describe rocks with visible crystal particles. Mostly used for carbonate and granitic rocks.

cyclothem – Alternating stratigraphic layers of marine and non-marine sediments with a basal sandstone, middle coal, and a top shale. Cyclothems are similar to a sequence, except they represent smaller units of time.

deformation – The process by which rocks undergo physical alteration through time. Often converting horizontal deposits into anticlines, synclines, domes, or basins. The process is driven by tectonics.

delta - A deposit of sediment, usually silt and clay, that forms where a stream enters a standing body of water such as a lake or ocean. When viewed from above they often appear fan or "D" shaped.

deposit – The direct settling of particles from suspension of a medium (air, water, ice, or from Earth's interior) or from direct chemical precipitation from a medium (usually water).

Devonian – The fourth geological period/system of the Paleozoic Era of the Phanerozoic Eon.

diamicton – Sediment that consists of a wide range of non-sorted to poorly sorted sediments such as sand or larger particle sizes that are suspended in a mud matrix that does not suggest a specific environment in which it formed. It is usually interchangeable with the word till.

dike – Describes a tabular or sheet like intrusion that cuts vertically or near vertical, through, and across beds of existing rock. Typically igneous in origin, but can be sedimentary in origin.

dip - The angle that a rock unit, relative to horizontal, of a planar feature, such as fault or bedding plane. The angle is measured downslope and perpendicular to the strike.

divergent boundary – Plate boundary where new crust is being created and plates are moving apart. They are often reflected on the surface as mid-ocean ridges and rift zones.

dolomite – A carbonate mineral composed of calcium magnesium carbonate, closely related to calcite.

dolostone – A sedimentary carbonate rock comprised predominately of the mineral dolomite. Often contains the same physical characteristics of limestone.

dome - An uplift that is round or elliptical in map view with beds dipping away in all directions from a central point. Older rocks are towards the center.

epicenter – The point at the Earth's surface directly above the focus of an earthquake.

erosion - A general term applied to the wearing away and movement of earth materials by gravity, wind, water, ice, chemical, or biological mechanisms.

facies - The lithological variation within a rock unit due to a change in depositional environment. Also an informal unit in stratigraphy.

fault –A break or fracture in rock along which measurable movement has occurred.

feldspar – A family of silicate minerals that originate in igneous rocks, also a major component in non-arenite sandstones.

fine grained – A term used to describe particles of rocks that are not visible with the naked eye.

fine grained sand– Particles of rock between 0.125 and 0.250 millimeters in diameter.

focus – The point within the Earth where an earthquake actually occurs.

fold – Describes layers of rock which have been bended or buckled from their original depositional position that are not faulted.

formation - A laterally continuous, somewhat homogenous rock unit with a distinctive set of characteristics that make it possible to recognize and map, from one outcrop or well to another. The basic formal rock unit in stratigraphy.

fracture – A local break along a plane in a geologic unit in which no measurable movement has occurred.

galena (mineral)– A gray metallic, heavy mineral formed predominately of lead and sulfide.

geology – The study of earth's physical structure, substance, history and the processes that act on it.

geosol – A body of sediment or rock composed of one or more soil horizons, usually applies to Quaternary soils.

glauconite – A soft mineral in the mica group. It colors sedimentary rocks green to dark green. It forms only in marine environments. Although it can occur in carbonate rock and shale, it is often associated with sandstones. The term green sand is often applied to sandstone rich in glauconite.

gneiss – A coarse grained, metamorphic rock that has a striped appearance caused by light-colored bands of granular minerals alternating with darker bands of platy or flaky minerals.

granite – A gray or pink, coarse-grained igneous rock that is made up mostly of plagioclase, potassium feldspar and quartz; but which may also contain mica or hornblende.

groundwater - Water that exists below the water table in an aquifer. Groundwater moves slowly in the between grains and fractures beneath the surface. It can also flow in underground streams.

growth fault – A type of normal fault that forms during sedimentation and typically has thicker strata on the downthrown (hanging) wall than the upthrown (foot) wall.

hardground – A term used to describe mineralization on a bedding surface that developed during a time of non-deposition.

hot spot – A large upwelling of magma from deep in the mantle, not along a plate boundary.

hydrothermal - Pertaining to the movement of hot water, the actions of hot water, or the deposits produced by the actions of hot water.

ice age – A period in Earth's history where continental sized glaciers have significantly covered the land.

igneous rock - A rock formed by the crystallization of magma or lava or by the ash that was deposited during one or more eruptions.

Illinois Episode (Illinoian) – A subdivision of the Quaternary during when sediments comprising the Illinoian glacial cycle in North America.

inland sea (continental seaway) – A shallow sea that covers central areas of continents during periods of high sea level that result in marine transgressions, also called epeiric seas.

interlobe – An area between separate, yet touching glaciers.

joint - A non-planar fracture in rock along which there has been no displacement.

Jurassic – The second geological period/system of the Mesozoic Era of the Phanerozoic Eon.

kame – A roughly round shaped hill of sand and gravel deposited at the front of a glacier as a debris fan or under a glacier.

Kansan – The youngest of the largely abandoned term used to describe pre-Illinois glacial episodes in North America.

kettle lake – A lake formed when an isolated segment of a glacier melts in the subsurface, forming a depression that becomes filled with water.

knob - A small rounded or elliptical hilltop that stands out from the surrounding hills.

limestone - A sedimentary rock consisting of at least 50% calcium carbonate ($CaCO_2$) by weight.

lithification – The physical process of turning sediment into rock.

lithology – The physical characteristics of a geologic unit such as structure, texture, composition, grain size, and color.

lithosphere - The outer layer of the Earth, that behaves rigidly. It consists of the ocean crust, the continental crust, and the upper part of the mantle. At its thickest, it is about 100 miles thick.

lobe – A tongue like projection from a continental glacier's main mass.

loess – Windblown deposits, usually of silt sized particles.

macro life – Multicellular life larger than microscopic in size.

magma – Molten rock material that occurs below Earth's surface.

mantle – A major subdivision of Earth's internal structure that exists in a semi-solid state. Located between the bottom of the crust and overlying the core. It constitutes about 83% of the planet's mass. Except for the lithosphere, it is plastic flowing rock. Also a term used to describe a thin layer of material over another type of material.

marble – A non-foliated metamorphic rock that is produced from the metamorphism of carbonate sedimentary rocks.

marine – Relating to the ocean and the activities that occur within.

meander – A natural bend in a stream or river that is "U" shaped.

member – A formal stratigraphic term used to classify small groups of similar rock in a localized area. It stratigraphy it is one rank lower than the formation.

Mesozoic – An era of geologic history that encompasses the Triassic, Jurassic, and Cretaceous Periods/Systems.

mica – A group of chemically and physically related minerals containing aluminum silicate, common in all three rock types.

micaceous – A term used to describe a rock containing significant amounts of one or more minerals of the mica group. In sandstones it tends to occur in small silvery flakes that look like glitter.

mineral – A solid, natural occurring substance, with a preferred molecular structure. They often form crystals.

Mississippian – The lower geologic subdivision of the Carboniferous Period/System.

mudstone – A sedimentary rock composed of clay and silt-size particles usually lacking defined bedding or lamination. Also used to describe sandy clays.

Nebraskan – The oldest of the two largely abandoned terms used to describe pre-Illinois glacial episodes in North America.

normal fault – A fault with net vertical movement that has occurred along a break that has formed a planar surface. The block above the fault has moved down relative to the block below the fault.

Ordovician – The second geological period/system of the Paleozoic Era of the Phanerozoic Eon.

oolite – A small sphere of carbonate or iron minerals, no more than a few millimeters in diameter and with a concentric internal structure . They primarily form by inorganic precipitation of calcium carbonate in very thin layers around a grain of sand or a particle of shell or coral.

outcrop – An exposure of a geologic unit(s) that is relatively free of vegetation. Outcrops can be formed naturally (streams and moving water) or by human action (road, rail, and building cuts).

oxbow – A short lived, crescent-shaped lake that forms when a meandering stream abandons a meander for a new course. Course changes frequently occur during flood events when overbank waters erode a new channel.

paleosol – An ancient soil that reflects exposure to the surface. Most do not exhibit direct organic activity but are largely chemically weathered surfaces. Used to describe pre-Quaternary soils.

Paleozoic – An era of geologic history that encompasses the Cambrian, Ordovician, Silurian, Devonian, Carboniferous, and Permian Periods/Systems.

Pennsylvanian – The upper geologic subdivision of the Carboniferous Period/System.

Period – A formal division of geologic time.

Permian – The sixth geological period/system of the Paleozoic Era of the Phanerozoic Eon.

Phanerozoic – A geologic eon beginning at the start of the Cambrian and continuing through today.

Pleistocene – The first epoch of the Quaternary Period, in which the all of the recent ice ages occurred.

plate – A physical ridged area of the Earth's crust, bordered by spreading, collision, and subduction zones. They can be composed of ocean crust, continental crust, or a combination of both.

pluton – An igneous rock formed deep beneath the surface of the Earth by consolidation of magma.

porphyritic – A texture term used to describe rocks containing distinct embedded crystals or crystalline particles.

porous – A term used to describe rocks containing numerous visible void space or pores.

Precambrian – A division of geologic time that encompasses all of Earth's history from its formation up to the beginning of the Phanerozoic Eon, encompassing ~88% of the Earth's history.

Pre-Illinois Episode (Pre-Illinoian) – Used by geologists to refer to the glacial advances and retreats before the Illinois Episode. It includes the old Nebraskan and Kansan glacial cycles.

quartz – A mineral composed entirely of silica (SiO_2), it is the most common rock-forming minerals at the Earth's surface.

quartzite – Usually a crystalline metamorphic rock formed by the alteration of quartz rich sandstone by heat and pressure, also can be of igneous or contact metamorphic in origin.

Quaternary – The most recent geological period/system of the Cenozoic Era of the Phanerozoic Eon.

reverse fault – A fault with vertical movement and an inclined fault plane. The block above the fault has moved upwards relative to the block below the fault.

road cut – A manmade cut in Earth's surface for the purpose of building a road.

rock – A hard aggregate of one or more minerals.

sand – A sedimentary rock composed of sand-sized particles (0.062 to 2 millimeters in diameter). Usually comprised of quartz but can contain feldspar, lithic fragments, and impurities such as silt and clay.

sandstone – A sedimentary rock composed of sand.

sediment – A loose, unconsolidated deposit of natural weathered debris, chemical precipitates, or biological debris that accumulates on Earth's surface.

sedimentary rock – A rock formed from the consolidation of sediment, usually in layered deposits or debris flows.

sequence – A series of rocks deposited in a marine transgressive to regressive environment. They are separated by unconformities. The term usually applies to sedimentary rocks but has also been used for igneous rocks.

shale – A clastic sedimentary rock that is made up of clay-size (less than 0.004 millimeter in diameter) particles. It typically breaks into thin flat pieces and usually laminated.

silt – A clastic sedimentary rock that forms from silt-size (between 0.004 and 0.062 millimeter diameter) particles.

Silurian – The third geological period/system of the Paleozoic Era of the Phanerozoic Eon.

Sloss sequence – Also known as a cratonic sequence, was proposed by Lawrence Sloss in 1963 to name transgressional seas that covered land on a continental scale.

sorting – The distribution of grain sizes within a rock. The more grains that are all the same size, the better sorted the rock is. A rock with similar grain size throughout is well sorted. A rock with a large variety of grain sizes is poorly sorted.

stream – The term is a matter of perspective but is usually used to refer to a small, narrow river. Also, a generic term for ancient rivers.

strike – A straight line that connects two points of the same elevation on a planar surface. It is always perpendicular to dip.

strike-slip fault – A fault on which net horizontal (instead of vertical) displacement occurs, typically caused by shear stress.

stromatolite – A carbonate or silica mat or mound-shaped fossil that forms from the repetitious layering of algal mat covered by trapped sediment particles. The oldest form of life recognizable in the field. They still exist today.

supercontinent – A large landmass that forms from the convergence of multiple continents, caused by Plate Tectonics.

syncline – An open, trough-shaped fold with youngest strata in the center.

System – A set of rocks deposited within a defined timeframe, equivalent to a Period.

tectonic(s) – The processes that move and deform Earth's crust.

Tertiary – The first geological period/system of the Cenozoic Era of the Phanerozoic Eon. The term was abandoned and what was the Tertiary has been divided into the Paleogene (23 to 66 million years ago) and the Neogene (2.6 to 23 million years ago).

tongue – A wedge shaped geologic unit in between another unit. A formal division of stratigraphy below the rank of formation.

thrust fault - A type of reverse fault that has a dip of less than 45°.

Triassic – The first geological period/system of the Mesozoic Era of the Phanerozoic Eon.

tufa – A variety of limestone formed by the precipitation of carbonate minerals from ambient temperature bodies of water, usually from groundwater and springs. It is also known as travertine.

till – An unsorted sediment deposited directly by a glacier and not reworked by melt water. Till is a form of diamicton.

unconformity – A contact between two rock units of different ages in which there is a span of time missing from the rock record.

vesicle/vesicular – A small and usually spherical cavity in a rock or mineral, formed by expansion of a gas or vapor before the enclosing body solidified. Also the tiny spaces in between fossils of a carbonate rock.

vug/vuggy - A small cavity in a rock. It is a larger version of a vesicle. They are often contain geodes.

weathering – A chemical or mechanical process, which breaks down rocks to smaller pieces and eventually deposited as sediment.

Wisconsin Episode – A subdivision of the Quaternary during when sediments comprising the Wisconsin glacial cycle in North America. It is the most recent of all glacial cycles.

Additional Reading

There is a great multitude of information on the geology of Illinois. The Illinois State Geological Survey now has all of its geologic maps and many of publications on its website to download and view for free. The Illinois State Museum has publications on fossils and the paleo-cultures of Illinois. The Midwest Institute of Geosciences and Engineering does not strictly focus on Illinois, but has many publications and maps pertaining to the Prairie State. Listed below are just a few of the sources of great information on the geology of Illinois.

Books:

Geology of Illinois (2010)

Geology Underfoot in Illinois (1997)

Guide to the Illinois Caverns State Natural Area (2004)

Richardson's Guide to the Fossil Fauna of Mazon Creek. (1997)

Websites:

Illinois State Geological Survey: www.isgs.illinois.edu

Illinois State Museum: www.museum.state.il.us

Midwest Institute of Geosciences and Engineering: www.mige-web.org

References

Atherton, E. Department of Registration and Education, Division of the State Geological Survey. (1947). *Some chester outcrop and subsurface sections in southeastern Illinois* (Circular No. 144). Urbana, Illinois. 122-131.

Barrows, H. H. (1910). *Geography of the middle Illinois valley*. (Vol. 15). Urbana, Illinois: State Geological Survey.

Baxter, J. Department of Registration and Education, Division of the State Geological Survey. (1960). *Salem limestone in southwestern Illinois* (Circular No. 284). Urbana, Illinois:

Bell, A. H. Department of Registration and Education, Division of the State Geological Survey. (1941). *Role of fundamental geologic principles in the opening of the Illinois basin* (Circular No. 75). Urbana, Illinois:

Bell, A. H. Department of Registration and Education, Division of the State Geological Survey. (1961). *Underground storage of natural gas in Illinois* (Circular No. 318). Urbana, Illinois:

Berg, R. C., Kempton, J. P., Follmer, L. R., & McKenna, D. P. Illinois Department of Energy and Natural Resources, Illinois State Geological Survey. (1985). *Illinoian and wisconsinan stratigraphy and environments in northern illinois: the altonian revised* (Guidebook Series No. 19). Urbana, Illinois:

Berg, R. C., McKay III, E. D., Goble, R. J., & Wang, H. Illinois Department of Natural Resources, Illinois State Geological Survey. (2013). *age of the winnebago formation of north-central illinois as determined by optically stimulated luminescence dating* (circular no. 580). Champaign, IL:

Bretz, J. H. Department of Registration and Education, Division of the State Geological Survey. (1939). *Geology of the chicago region* (Bulletin No. 65, Part I). Urbana, Illinois:

Bretz, J. H. Department of Registration and Education, Division of the State Geological Survey. (1955). *Geology of the chicago region* (Bulletin No. 65, Part II). Urbana, Illinois:

Bretz, J. H., & Harris, Jr., S. E. Department of Registration and Education, Illinois State Geological Survey. (1961). *Caves of illinois* (Report of Investigations 215). Urbana, Illinois:

Bristol, H. M., & Howard, R. H. Department of Registration and Education, Illinois State Geological Survey. (1971). *paleogeologic map of the sub-pennsylvanian chesterian (upper mississippian) surface in the illinois basin* (Circular No. 458). Urbana, Illinois:

Chrzastowski, M. J. Illinois State Geological Society, (2009). *The chicago river - a legacy of glacial and coastal processes* (Guidebook Series No. 37). Urbana, Illinois:

Cluff, R. M., Reinbold, M. L., & Lineback, J. A. Department of Registration and Education, Illinois State Geological Survey. (1981). *The new albany shale group of illinois* (Circular No. 518). Urbana, Illinois:

Curry, B. B., Graese, A. M., Vaiden, R. C., Bauer, R. A., Schumaher, D. A., Norton, K. A., Dixon, Jr., W. G., & Reed, P. C. Illinois Department of Energy and Natural Resources, Illinois State Geological Society. (1988). *Geological-geotechnical studies for siting the superconducting super collider in illinois: results of the 1986 test drilling program* (Environmental Gology Notes No. 122). Urbana, Illinois:

Dey, W. S., Davis, A. M., Curry, B. B., Keefer, D. A., & Abert, C. C. Illinois State Geological Society, (2007). *Kane county water resources investigations; final report on geologic investigations* (ISGS Open File Series 2007-7). Champaign, IL:

Ekblaw, G. E. Department of Registration and Education, Division of the State Geological Survey. (1938). *Kankakee arch in illinois* (Circular No. 40). Urbana, Illinois.

Emrich, G. H. Department of Registration and Education, Illinois State Geological Survey. (1966). *Ironton and galesville (cambrian) sandstones in illinois and adjacent areas* (Circular No. 403). Urbana, Illinois:

Frankie, W. T., Kolata, D. R., & Berg, R. C. Illinois Department of Energy and Natural Resources, Illinois State Geological Survey. (1999). *Guide to the geology of the rock cut state park and rockford area, winnebago county, illinois* (Field Trip Guidebook 1999C, 1999D). Urbana, Illinois:

Frankie, W. T., & Nelson, R. S. Illinois Department of Natural Resources, Illinois State Geological Survey. (2002). *Guide to the geology of the apple river canyon state park and surrounding area of northeastern jo daviess county, illinois.* Urbana, Illinois:

Frye, J. C., Willman, H. B., & Glass, H. D. Department of Registration and Education, Illinois State Geological Survey. (1964). *Cretaceous deposits and the illinoian glacial boundary in western illinois* (Circular No. 364). Urbana, Illinois:

Graeses, A. M. Department of Registration and Education, Illinois State Geological Survey. (1991). *facies analysis of the ordovician maquoketa group and adjacent strata in kane county, northeastern illinois* (Circular No. 547). Urbana, Illinois:

Grimley, D. A., & Phillips, A. C. Illinois State Geological Society, (2011). r*idges, mounds and valleys: glacial-interglacial history of the kaskaskia basin, southwestern illinois* (ISGS Open File Series 2011-1). Champaign, IL:

Hansel, A. K., Berg, R. C., Philips, A. C., & Gutowski, V. G. Illinois Department of Energy and Natural Resources, Illinois State Geological Survey. (1999). *glacial sediments, landforms, paleosols, and a 20,000 year old forest bed in east-central illinois* (Guidebook Series No. 26). Urbana, Illinois:

Hansel, A. K., & Johnson, W. H. Department of Registration and Education, Illinois State Geological Survey. (1996). Wedron and mason groups: Lithostratigraphic reclassification of deposits of the wisconsin episode, lake michigan lobe area (Bulletin No. 104). Urbana, Illinois:

Hensel, B. Illinois Department of Energy and Natural Resources, Illinois State Geological Survey. (1992). *Natural recharge of groundwater in illinois* (Environmental Gology Notes No. 143). Urbana, Illinois:

Hughes, G. M., Kraatz, P., & Landon, R. A. Department of Registration and Education, Illinois State Geological Survey. (1966). *Bedrock aquifers of northeastern illinois* (Circular No. 406). Urbana, Illinois:

Johnson, W. H., Follmer, L., Gross, D. L., & Jacobs, A. M. Illinois State Geological Survey, (1972). *Pleistocene stratigraphy of east-central illinois* (Guidebook Series No. 9). Urbana, Illinois:

Johnson, W. H., Hansel, A. K., Bettis III, E. A., Karrow, P. F., Larson, G. J., Lowell, T. V., & Schneider, A. F. Illinois Department of Natural Resources, Illinois State Geological Survey. (1996). *late quaternary temporal and event classifications, great lakes region, north america* (Reprint series 1997B). Champaign, IL:

Kimball Brown, M. (1975). The Zimmerman site: Further excavations at the grand village of kaskaskia. *Illinois State Museum: Reports of Investigation*, (32),

King, J. E. (1982). Fossils. *Illinois State Museum: Story of Illinois Series*, (14),

Kolata, D. R. Illinois State Geological Survey, (1991). *Tippecanoe i subsequence; middle and upper ordovician series* (Reprint series 1991-T5). Champaign, IL:

Kolata, D. R. Illinois State Geological Survey, (1991). *Tippecanoe sequence overview; middle ordovician series through lower devonian series* (Reprint series 1991-T4). Champaign, IL:

Kolata, D. R., & Buschbach, T. C. Department of Registration and Education, Illinois State Geological Survey. (1976). *plum river fault zone of northwestern illinois* (Circular No. 491). Urbana, Illinois:

Kolata, D. R., Buschbach, T. C., & Treworgy, J. D. Department of Registration and Education, Illinois State Geological Survey. (1978). *The sandwich fault zone of northern illinois* (Circular No. 505). Urbana, Illinois:

Kolata, D. R., & Nimz, C. K. (2010). *Geology of Illinois*. Champaign, IL: University of Illinois Board of Trustees.

Kolata, D. R., Treworgy, J. D., & Masters, J. M. Department of Registration and Education, Illinois State Geological Survey. (1981). *Structural framework of the Mississippi embayment of southern Illinois.* (Circular No. 516). Urbana, Illinois:

Leetaru, H. E., Sargent, M. L., & Kolata, D. R. Illinois Department of Natural Resources, Illinois State Geological Survey. (2004). *Geologic atlas of cook county for planning purposes*. Champaign, IL:

Mast, R. F., Ruch, R. R., & Meents, W. F. Department of Registration and Education, Illinois State Geological Survey. (1973). *vanadium in devonian, silurian, and ordovician crude oils of illinois* (Circular No. 483). Urbana, Illinois:

McGinnis, L. D. Department of Registration and Education, Illinois State Geological Survey. (1966). *Crustal tectonics and precambrian basement in northeastern illinois* (Report of Investigations 219). Urbana, Illinois:

Mehnert, E. Illinois Department of Natural Resources, Illinois State Geological Survey. (2010). *Groundwater flow modeling as a tool to understand watershed geology: blackberry creek watershed, kane and kendall counties, illinois* (circular no. 576). Champaign, IL:

Mehnert, E., Hackley, K. C., Larson, T. H., Panno, S. V., Pugin, A., Wehmann, H. A., Holm, T. R., & Roadcap, G. S. Illinois State Geological Society, Illinois State Water Survey. (2004). *The mahomet aquifer: recent advances in our knowledge* (ISGS Open File Series 2004-16). Champaign, IL

Melhorn, W. N. Illinois Department of Energy and Natural Resources, Illinois State Geological Survey. (1991). *Tippecanoe i subsequence; middle and upper ordovician series* (Reprint series 1991L). Champaign, IL:

Mikulic, D. G., & Kluessendori, J. Illinois Department of Energy and Natural Resources, Illinois State Geological Survey. (1999). *the classic silurian reefs of the chicago area* (Guidebook Series No. 29). Urbana, Illinois:

Nelson, W. J. Department of Natural Resources, Illinois State Geological Survey. (2002). *Sequence stratigraphy of the lower chesterian (mississippian) strata of the illinois basin* (Bulletin No. 107). Champaign, Illinois:

Nelson, W. J. Department of Registration and Education, Illinois State Geological Survey. (1995). *Structural features in illinois* (Bulletin No. 100). Urbana, Illinois:

Nelson, W. J., & Lumm, D. K. Department of Registration and Education, Illinois State Geological Survey. (1987). *Structural geology of southeastern Illinois and vicinity.* (Circular No. 538). Urbana, Illinois:

Nelson, W. J., & Welbel, C. P. Department of Registration and Education, Illinois State Geological Survey. (1996). *Geology of the lick creek quadrangle johnson, union, and williamson counties, southern illinois* (Bulletin No. 103). Urbana, Illinois.

Potter, P. E. Department of Registration and Education, Division of the State Geological Survey. (1962). *Late mississippian sandstones of illinois* (Circular No. 340). Urbana, Illinois:

Potter, P. E. Illinois State Geological Survey, (1963). *Late paleozoic sandstones of the illinois basin* (Report of Investigations 217). Urbana, Illinois:

Reed, P. C. Illinois State Geological Society, (1974). *Data from controlled drilling program in boone and dekalb counties, illinois* (Environmental Geology Notes No. 77). Urbana, Illinois:

Reinertsen, D. L. Illinois Department of Energy and Natural Resources, Illinois State Geological Survey. (1992). *Guide to the geology of the galena area* (Field Trip Guidebook 1992B). Urbana, Illinois:

Reinertsen, D. L., Jacobson, R. J., Killey, M. M., Nelson, R. S., & Reed, P. C. R. (1993). *Guide to the geology of the Lewistown-spoon river area Fulton County, Illinois*. Champaign, IL: Illinois State Geological Survey.

Reynolds, R. L., Goldhaber, M. B., & Snee, L. W. (1997). Paleomagnetic and 40ar/39ar results from the Permian Downeys Bluff Sill - evidence for Permian igneous activity at Hicks Dome, southern Illinois Basin. In R. W. Scott Jr. (Ed.), *Evolution of Sedimentary Basins - Illinois Basin*. Denver, CO: U.S. Department of the Interior.

Rogers, C., (2019). *The mysterious Tully Monster just got more mysterious*. Live Science News Release

Sargent, M. L. Illinois State Geological Survey, (1991). *Sauk sequence; cambrian system through lower ordovician series.* Champaign, IL:

Shrode, R. S. Department of Registration and Education, Division of the State Geological Survey. (1948). *Unusual oolite grains from the ste. genevieve limestone* (Circular No. 144). Urbana, Illinois. 140-144.

Stohr, C. J., Petras, J., Mikulic, D. G., & Thomason, J. (2011). Stereophotographic measurement of joint and bedding orientation at thornton quarry, illinois. In Illinois State Geological Survey.

Stumpf, A. J., Hansel, A. K., & Barnhardt, M. L. (2003). *Geologic mapping of glacial drift aquifers in the greater Chicago area of Illinois*. Champaign, IL: Illinois State Geological Survey.

Swann, D. H., Lineback, J. A., & Frund, E. Department of Registration and Education, Illinois State Geological Survey. (1965). *The borden siltstone (mississippian) delta in southwestern illinois* (Circular No. 386). Urbana, Illinois:

Templeton, J. S., Graf, D. L., Horberg, L., & Workman, L. E. Department of Registration and Education, Division of the State Geological Survey. (1951). *Short papers on geologic subjects* (Circular No. 170). Urbana, Illinois.

Templeton, J. S., & Willman, H. B. Department of Registration and Education, Illinois State Geological Survey. (1963). *Champlainian series (middle ordovician) in illinois* (Bulletin No. 89). Urbana, Illinois.

Treworgy, J. D. Department of Registration and Education, Illinois State Geological Survey. (1988). *The illinois basin - a tidally and tectonically influenced ramp during mid-chesterian time* (Circular No. 544). Urbana, Illinois:

Tri-State Committee on Correlation of the Pennsylvanian System in the Illinois Basin. Illinois State Geological Survey, Indiana. Geological Survey, & Kentucky Geological Survey, (2001). *Toward a more uniform stratigraphic nomenclature for rock units (formations and groups) of the Pennsylvanian system in the Illinois basin*. Illinois Basin Consortium.

Udden, J. A. (1914). *Some deep borings in illinois*. (Vol. 24). Urbana, Illinois: Illinois State Geological Survey.

Vaiden, R. C., Smith, E. C., & Larson, T. H. Illinois Department of Natural Resources, Illinois State Geological Survey. (2004). *groundwater geology of dekalb county, illinois with emphasis on the troy bedrock valley* (Circular No. 563). Urbana, Illinois:

Visocky, A. P., Sherrill, M. G., & Cartwright, K. Illinois Department of Energy and Natural Resources, Illinois State Geological Survey, Illinois State Water Survey. (1985). *Geology, hydrology, and water quality of the cambrian and ordovician systems in northern illinois* (Cooperative Groundwater Report 10). Urbana, Illinois:

Walter, P. (1963). Transactions of the Illinois state academy of science. In *Transactions of the Illinois State Academy of Science* (Vol. 56, pp. 59-67). Urbana, Illinois: Illinois State Geological Survey.

Webb, N. D., Grimley, D. A., Phillips, A. C., & Fouke, B. W. (2012). Origin of glacial ridges in the Kaskaskia Sublobe, southwestern Illinois, USA. *Quaternary Research, (78), 341-352.*

Weller, J. M. Department of Registration and Education, Division of the State Geological Survey. (1943). *Rhythms in upper pennsylvanian cyclothems* (Circular No. 92). Urbana, Illinois:

Whitaker, S. T. Illinois Department of Energy and Natural Resources, Illinois State Geological Survey. (1988). *Ramp-platform model for silurian pinnacle reef distribution in the illinois basin*. Urbana, Illinois:

Whiting, L. L., & Stevenson, D. L. Department of Registration and Education, Illinois State Geological Survey. (1965). *The sangamon arch* (Circular No. 383). Urbana, Illinois.

Wiggers, R., Mountain Press Publishing Company. (1997). *geology underfoot in illinois*. Missoula Montana

Willman, H. B. Department of Registration and Education, Illinois State Geological Survey. (1973). *rock stratigraphy of the silurian system in northeastern and northwestern illinois* (Circular No. 479). Urbana, Illinois:

Willman, H. B. Department of Registration and Education, Illinois State Geological Survey. (1971). *summary of the geology of the chicago area* (Circular No. 460). Urbana, Illinois:

Willman, H. B., Atherton, E., Buschbach, T. C., Collinson, C., Frye, J. C., Hopkins, M. E., Lineback, J. A., & Simon, J. A. Department of Registration and Education, Illinois State Geological Survey. (1975). *Handbook of illinois stratigraphy* (Bulletin No. 95). Urbana, Illinois.

Willman, H. B., & Frye, J. C. Department of Registration and Education, Illinois State Geological Survey. (1969). *high-level glacial outwash in the driftless area of northwestern illinois* (Circular No. 440). Urbana, Illinois:

Willman, H. B., & Frye, J. C. Department of Registration and Education, Illinois State Geological Survey. (1970). *Pleistocene stratigraphy of illinois* (Bulletin No. 94). Urbana, Illinois:

Willman, H. B., & Kolata, D. R. Department of Registration and Education, Illinois State Geological Survey. (1978). *the platteville and galena groups in northern illinois* (Circular No. 502). Urbana, Illinois:

Willman, H. B., Lowenstam, H. A., & Workman, L. E. Illinois State Geological Survey. (1950). *Guidebook: Field conference on niagaran reefs in the chicago region*. Urbana, Illinois:

Willman, H. B., & Payne, J. N. Department of Registration and Education, Division of the State Geological Survey. (1943). *Early ordovician strata along fox river in northern illinois* (Circular 100). Urbana, Illinois:

Willman, H. B., & Payne, J. N. Department of Registration and Education, Division of the State Geological Survey. (1942). *Geology and mineral resources of the marseilles, ottawa, and streator quadrangles* (Bulletin No. 66). Urbana, Illinois:

Workman, L. E. Department of Registration and Education, Division of the State Geological Survey. (1938). *The preglacial rock river valley as a source of groundwater for rockford* (Circular No. 36). Urbana, Illinois:

Maps

Barnhardt, M.L., Institute of Natural Resource Sustainability, Illinois State Geological Survey (2009), *surficial geology of zion quadrangle, lake county, illinois, and kenosha county, wisconsin*, Statemap Zion-SG

Baumann, S.D.J., Midwest Institute of Geosciences and Engineering (2011), *surficial geologic map of castle rock state park & lowden miller state forest, ogle county, Illinois, U.S.A.*, M-092011-1A

Baumann, S.D.J., Midwest Institute of Geosciences and Engineering (2011), *surficial geologic map of keepataw forest preserve and lemont area, will, cook, and dupage counties, Illinois, U.S.A*, M-102011-1A

Baumann, S.D.J., Midwest Institute of Geosciences and Engineering (2013), *surficial geology of the romeoville quadrangle, part of cook, dupage, and will counties, Illinois*, United States, M-112013-3A

Baumann, S.D.J., *Midwest Institute of Geosciences and Engineering (2014), surficial geologic map of the Stockton quadrangle, jo daviess county, Illinois, United States*, M-102014-1B

Denny, F.B., King, B., Mulvaney-Norris, J., Malone, D., Institute of Natural Resource Sustainability, Illinois State Geological Survey (2010), *bedrock geology of kabers ridge quadrangle, hardin, gallatin, and saline counties, Illinois*, IGQ Kabers Ridge-BG

Denny, F.B., Nelson, W.J., Munson, E., Devera, J.A., Amos, D.H., Prairie Research Institute, Illinois State Geological Survey (2011), *bedrock geology of rosiclare quadrangle, hardin county, Illinois*, Statemap Rosiclare-BG

Harrison, R.W., U.S. Department of the Interior, U.S. Geological Survey (1999), *geologic map of the thebes quadrangle, Illinois and Missouri*, MAP GQ-1779

Jacobson, R.J., Weibel, C.P., Illinois Department of Natural Resources, Illinois State Geological Survey (1993), *geologic map of the makanda quadrangle, Jackson, union, and willamson counties, Illinois*, IGQ-11

Nelson, W.J., Devera, J.A., Illinois Department of Natural Resources, Illinois State Geological Survey (2007), *bedrock geology of mt. pleasant quadrangle, union and Johnson counties, Illinois*, IGQ Mt. Pleasant-BG

Disclaimer

The vast majority of the photos throughout Volumes I and II were taken in 2011-2013 during field visits. At the time this second edition is going to print, one of the authors still lives in Illinois. Sometimes sites change. Roadcuts get covered or disappear due to road expansion, parks get upgraded, floods damage outcrops, people build new structures, etc. Every reasonable effort was made to make sure all the sites visited in Volume II are up to date. If changes have occurred, we try to mention them. We are not infallible, and with the primary author (Steven Baumann) no longer living in Illinois, there can be things we missed.

www.ingramcontent.com/pod-product-compliance
Lightning Source LLC
Chambersburg PA
CBHW051152220526
45473CB00003B/742